University of Tehran Science and Humanities Series

Series Editor

Central Acquisitions Office, University of Tehran, Tehran, Iran

The *University of Tehran Science and Humanities Series* seeks to publish a broad portfolio of scientific books, basically aiming at scientists, researchers, students and professionals. The series includes peer-reviewed monographs, edited volumes, textbooks, and conference proceedings. It covers a wide range of scientific disciplines including, but not limited to Humanities, Social Sciences and Natural Sciences.

More information about this series at http://www.springer.com/series/14538

Ali Asghar Adibi

Collage: A Process in Architectural Design

Springer

Ali Asghar Adibi
University of Tehran
Tehran, Iran

Translated by
Ali Yaser Jafari
University of Tehran
Tehran, Iran

Reihaneh Khorramrouei
University of Tehran
Tehran, Iran

ISSN 2367-1092 ISSN 2367-1106 (electronic)
University of Tehran Science and Humanities Series
ISBN 978-3-030-63797-2 ISBN 978-3-030-63795-8 (eBook)
https://doi.org/10.1007/978-3-030-63795-8

This Springer imprint is published by the registered company Springer Nature Switzerland AG
The registered company address is: Gewerbestrasse 11, 6330 Cham, Switzerland

A city is never seen as a totality, but as an aggregate of experiences, animated by use, by overlapping perspectives, changing light, sounds, and smells. Similarly, a single work of architecture is rarely experienced in its totality but as a series of partial views and synthesized experiences. Questions of meaning and understanding lie between the generating ideas, forms, and the nature and quality of perception.

In generally, "The world is a COLLAGE"

"JuhaniPallasmaa"

Foreword

World as Complexity and Multiplication

Although lots of valuable intellectual, scientific and industrial achievements came with the twenty-first century, modernism movement followed a linear goal that was not quite aligned with the predominant economical–political approaches in the world, and despite all its social claims, it can still be characterized as a theatrical artistic movement. Relying on logical action, essence is probably one of the important approaches of modernism and consequently moving architecture from a classical metaphysic position to a natural and empirical science position, and hence carrying a kind of pseudo-wisdom positivism inside. Along with it, multiplicity of the evolving world seems to be overshadowed, and some industrial countries known as first world pursuing their ulterior colonial goals and taking advantage of new combination of cold wars tend to dominate their own intellectual rules. In this way, linear process of architecture as a well-known principle has developed and human ascendancy has entered floating grounds and intends to establish conceptual universalities that are never considered in square equilibriums. Linear process of modernism has intensive vision intransigence which disregards many visible and invisible strata of contemporary world and under the name of "art" overrules the culture, tradition and history and replace the fact and comprehensiveness of social paradigm with abstract artistic patterns. Mottoes of absolute rules can only analyze some parts of the facts and the phenomenon, and the diminution is recorded and implemented as scientific salvation.

In which scientific-intellectual grounds has theatrical architecture taken its strong roots and how can a culture and a continent bring the world's philosophical theatrical unrestricted rules under their domination? Unlike apertures through which only a small portion of human cognition can be seen, looking at architecture is like looking at the universe. Moving pass the algorithms with mercurial expansion is an indicative of an apparent incongruent development which betokens the complexity of the development that needs to be further explored.

In liberation theory by Roland Barthes as a complementary ground, it is aimed to implement a multidimensional changing approach whose basis can receive various extremely complicated compounds in layers. Maintaining the liaison of various segments in order to reach a practical depth is one of the purposes of comprehensive outlook of diversities.

In collage, compared to other reductive methods, the span and dispersion of the process to increase the reception of diffused thoughts with the purpose of gaining expansion and vision multiplicity is discussed. Furthermore, unlike Descartes and Kant's rationalism in which logical reasoning for phenomena was supported, deconstructionism is based on the lack of decisiveness and the interpretation of the occurred event is embedded in its core approach. In spite of ostensible freedom of speech, the event aspect has the history, culture and memory in its heart and its abstraction functions only as a subject creator. In doing so, the visible mind and the invisible memory as the essentials of the imprinted creativity phenomenon and detection and reception of image also carry the seeds of cultural–historical linkage.

In recent decade, the affinity between science, philosophy and art has been noticeable in all phenomena and entering the various grounds has become necessity. The discoveries begin with the unknowns, and subconscious and intuition are among the topics that can merely be fertilized by an immense capacity of multiplicity. Where should we start the line of architecture to overcome the existing challenges and to dispose of the scholastic dogma? It is only through interaction and capacity augmentation of the multicipities and floating complexities that this question can be answered and probably diversity of the recitative and conceptual demonstration are reliant on it.

Transformation of dogmas is the first requirement to be considered; and establishing a principle under the name of science will result in diminishing the universal cognition in some parts. Science, art, society, philosophy and cognition have eternal affinity, and the line and discourse of architecture unavoidably have to pass these intersections of incidents and conflicts and endure its diverse conceptual difficulties.

Dr. Adibi has adopted this pluralistic approach in his layer exploration of collage whose instantaneous occurrence encompasses the unknown images of his freethinking mentality. Freedom is the first requirement of an artistic approach, thus opposing all forms of dogmas or the absolute law of process that by omitting the multicity creates an imperfect world.

For the last 15 years, we have witnessed all Ali Asghar Adibi's exertions to institute another stage of events which were sometimes impeded by some objections. Are these objections new forms of historical and stereotypical dogmas which have reared their ugly heads here again? We can deliberate.

Throughout history, there are numerous instances of human explorations. The method presented in this study can be considered as means to eliminate the old darkness, dogmas and empiric stereotypes and can be beneficial to all the people who seek to discover the latent mentalities.

In music, the polyphonic phoneme is enormously comparable to the extent of today's world visual communication whose effect in architecture is exposed as today's multifaceted conceptual opening.

There are many mysteries resided in collage. Like discovery of the floating particles looking for their totalities, anti-dogma and anti-stereotype method of collage can lead us to the visual and then intellectual freedom which is manifested in its shift in today's holistic politics and globalization. The verification of Adibi's work and effort by an accredited scientific institution such as M.I.T. University is the confirmation of this fact which is also an indicative of an expansive universal scientific–artistic communication imbedded in this invaluable and impressive message.

Adibi opines in the manner of "tendency for difference" that according to Derrida, in the world of difference, one must look for the conceptual multiplicities. While leaning towards Maulana Jalaluddin Balkhi's viewpoint, interpretation generating is so close to Deleuze's theorization and eventually to Iranians' convictional and original culture.

In the world of mysticism and spirituality, the multi-layered firmament is referred to as the ascension in which the revelation is achieved in phases and exceptional states of spirit, and the enlightenment of the fact is accompanied by spirituality and cognition, a transcendental art produced by deep existential worship. Looking at these works and sympathizing with this end will probably position us as an unbiased witness who is capable of observing the events in a novel manner and perhaps on account of that, we can look at the end of this collection of experiences and efforts delightedly. What is the mystery of the world? It is what we have not discovered yet…

Tehran, Iran
November 2007

Darab Diba

Preface of Translators

"Collage and montage" are quintessential techniques in modern and contemporary art and filmmaking. Collage combines pictorial motifs and fragments from disconnected origins into a new synthetic entity which casts new roles and meanings to the parts. It suggests new narratives, dialogues, juxtapositions and temporal durations. Its elements lead double-lives; the collaged ingredients are suspended between their originary essences and the new roles assigned to them by the poetic ensemble.

The idea of collage has also been a conscious and deliberate artistic method in architecture from Giulio Romano's Palazzo Te in Mantua to Le Corbusier's and Alvar Aalto's fusions of modernist and vernacular images in their architectural assemblages, all the way to Jean Nouvel's collaged walls of the Belfort Theatre and David Chipperfield's renovation of Neues Museum in Berlin.

Using "collage" among Iranian students in architecture studio is vast. In order to introduce the way these students use this technique to the English reader, we have chosen the valuable book *Collage: A Process in Architectural Design* by Ali Asghar Adibi to translate it in English, which we think provides a representative samples of design through collage on other culture. This book is originally collected and published in three chapters: "Collage History in Different Arts, Objectives and Steps to Make Collage Images" and the last chapter is "Two Experienced Examples". Chapter 1 deals with the origin of this technique and its usage in music, digital, architecture, etc. In Chap. 2, Dr. Adibi wrote about aim of advantages of collage, tools and materials and atmosphere in collage workshops.

In Chap. 3, two subjects of designing in Dr. Adibi and Dr. Farzin classes (Design of architect's house and Chizar cultural center) have been dealt with. In generally, further explanatory materials are located in this chapter. Each student express his/her experienced design and name it based on their feelings on collage works and skills, for instance "From inside", "Crossing the limits", "Hidden identity", "From dynamism to inertia", "The color of gray mind", etc. At the end, a summary of the whole book is represented.

Finally, we will be grateful for reports of any errors or omissions in this file; they will be corrected in the future editions.

First and foremost, we'd like to especially thank our teacher, Dr. Ali Asghar Adibi, provides the necessary time, guides us and translation of this book would not have come to fruition without his efforts.

We'd like to thank our parents for allowing us to follow our ambitions throughout our childhoods. My family, including our mothers, have always supported us throughout our careers, helped us to gather our precious pieces of life like collage, and we really appreciate it.

Last, but definitely not least, I would like to express my gratitude to the many people who saw me through this book; to all those who provided support, talked things over, read, wrote, offered and assisted in the editing, proofreading and design.

Eng. Ali Yaser Jafari
Ph.D. Candidate of Architecture
University of Tehran
Tehran, Iran

Eng. Reihaneh Khorramrouei
Ph.D. Candidate of Landscape Architecture
University of Tehran
Tehran, Iran

Author's Prologue

The present book is the result of three decades of study, research and teaching on the ground of creativity and self-fertility and their effects upon the development of architectural design. In this book with regard to this technique, initially, we will discover the space as the most fundamental and structural concept in architecture and then we will continue with the essence of primary aesthetics (which is the outcome of various innovative intuitive practices) which based on the effects of environmental, cultural, logical and functional factors will evolve and reach its final stage.

In this method, an effect based on images, sequence of mental creativity, objective interpretation of abstract images and making them transferrable to others is attained through three-dimensional simulation method whose result is the effect which has direct relationship with designer's depth of vision, thought and imagination.

Even though, extracting the space from some integrated images can be a challenge for any new art students, constant practice, interest and mutual trust between teacher and the student can be considered as some solutions to this problem. The first step in this method, hence, is summoning the courage to experience it and building trust between teacher and student.

Regretfully, the majority of art students who choose to pursue the art academically lack the artistic basis necessary to this field which can be blamed on pre-university education structure for giving excess emphasis on theoretic, mathematical and logical sciences, hence depriving them of artistic creativity and leading them to experience a difficult time adapting to an actual artistic environment. By making it their obligation to make this transition and adaptation less painful for students, the teachers will also become involved. The other problem needed to be addressed is the lack of possibility for the students to change their field of study in case of failure to make the adaptation. We need to acknowledge the fact that despite their tremendous effort to adapt to this field of study, only a few students possess the required qualities to succeed. This heterogeneity will cause some difficulties. However, there are some diligent and self-confident students who will ultimately manage to be a successful architect. This method which is brimful of creativity and innovation can be advantageous to these students.

In this collection, providing some overall explanation about this method, we have presented the architectural designs created by the students who all applied this method. It is also provided for them to express their personal opinions individually. Regarding the students' works, I have predominantly strived to retain the originality of the designs and maintain the freedom in the writings. I will never be to reciprocate the favour and the genuine kindness that all my friends and colleagues had towards me in creating this book, particularly, D.Eng. Ahmad Ali Farzin as my caring companion in guiding the students, D.Eng. Seyedeh Neda Ghazi Zadeh, Ms. Parastoo Eshrati (Eng.), Ms. Maryam Farhadi (Eng.) who kindly assisted me all the way through the writing and particularly edition of this book. The constructive role and accountability of students should also be appreciated, and I owe it to myself to pay my tribute to all of them.

The present book has introduced a new trend in teaching process and architectural design to our country's scientific and academic community. Acknowledging that it is not a flawless work, I look forward to receiving criticisms and recommendations from readers, researchers and scholars of the scientific community so that I could gladly implement them in the next edition.

Tehran, Iran Ali Asghar Adibi

Introduction

Human awareness of his environment starts with his birth or even before. He stares at his mother's eyes, looks, sees his cradle, people and the world around him and sometimes stares at something whose existence is unknown to him. He has a small world. As he grows up, he starts walking with his parents, his perception of the alley is not the same as his parents', for him, alley is a place to be covered: finding the answers to thousands of questions imprinted in his mind. He has never seen the places he is seeing. The world is like a big question to him.

He grows up and starts walking by himself. He alone travels; sees; learns. Inadvertently and unknowingly, he sees more and more and learns more and more. He passes through different stages of life carrying a large load of questions and answers which becomes heavier with the passing of time. This load is the one that forms his insight and shapes his personality. While expressing love towards something, he walks past something else indifferently. While accepting something enthusiastically, he rejects something else stubbornly. This insight formed in the cultural structure of society changes constantly, yet it exists.

Using this "insight", we can think, make decision and draw lines as a designer. Some questionable decision! Good or bad, right or wrong, these decisions are implemented.

But how do we draw a line? What process does something that flows instantly and immediately on the paper go through? Such questions might initially seem unanswerable. But without doubt "our creativity" plays a critical role in this process. Our creativity is something that is unique to us and is honed during our life.

Acknowledging that some people are "gifted artists", we also emphasize the fact that like our body, our mind needs to be trained too, or it will remain limited and that unique creativity won't be boosted.

One of the principles of creating a work of art is the visual accumulations of one's "invisible memory" and "visible memory". Most of the images we have in our mind are the repetition or the product of the existing facts process. But these existing facts and new work of art are located at both ends of the factors which embrace the overt and covert images.

New artwork and the hidden memory images ← We —— We —— We →, new effect and visible memory images which direction we are drawn to is directly dependent on our ability to discover "the hidden memory images" or the selection of "the visible memory images". In the process of creating the artwork, in the struggle between logic and imagination, consciousness and subconscious, imagination and subconscious are always victorious.

The less "key images" derived from a work of art are affected by logic and consciousness, the more successfully they are restated, thus all the points, lines, images shapes, spaces, etc., are formed in designer's mind and the recreation of them in external world is a transition from "subconscious to consciousness" which transfers the contexture and profoundness inspired by "key images" to the created spaces.

There are images in our "visible memory" whose volume is dependent on our amount of experiences. But invisible images are indicative of our other side of personality which are only exposed in special conditions. Designing the ideal spaces can serve as an opportunity to stimulate the designer's creativity. It is an effort to establish the communication with the infinite external world, to identify the unique aptitudes of the designer, to believe them, to build self-confidence, courage to create and ideate and finally to reach an ideal space which is unique and genuine. The space which is obtained via relying on facts and internal desires. What we are looking for is to discover a way to reveal these hidden images; the images which are dependent on "our" inside, insight formed during "our" history, "our" personal anguish and the visualized ideals in "our" subconscious. "Collage" is one of these ways.

Collage is a channel to the innermost recesses of "our" mind searching for the source of images which regardless of scale, form and space frames flow out of "our" subconscious and into the frame of each scale, form and space.

Contents

1 A Brief History of Collage 1
- 1.1 The Origins of This Method 1
- 1.2 Collage in Painting 2
- 1.3 Collage in Architecture 3
- 1.4 Wooden Collage 3
- 1.5 Collage in Music 4
- 1.6 Digital Collage 4
- 1.7 Literary Collage 4
- 1.8 Decoupage 4
- 1.9 Photomontage 4

2 Objectives and Steps to Make Collage Images 7
- 2.1 Advantages and Objectives of Collage 7
- 2.2 Tools and Materials and How to Use Them 9
- 2.3 Design Method in Collaging 10
- 2.4 Color Painting in Collage 14
- 2.5 Collage Workshop Mood 15

3 Two Cases of Experienced Examples 25
- 3.1 ■ Architect's Housing Design (The Second Home) 25
- 3.2 Axes Fracture 27
- 3.3 "Collage", a Novel Method in a Novel World 33
- 3.4 "Collage", a Novel Dialog 41
- 3.5 The Forgotten Details 45
- 3.6 Involvement with Tranquility 51
- 3.7 An Ideal Start 55
- 3.8 Intellectual Motion from Inside to Outside 59
- 3.9 The Sparkle of the Mind of "Collage" 63
- 3.10 From Inside 67
- 3.11 Searching for the Favorite Space 71
- 3.12 Recovering Spaces in the Hidden Corners of the Mind 77
- 3.13 Releasing from Red Lines 81
- 3.14 The Hidden Identity 85
- 3.15 Awakening Pictures in the Memory 87
- 3.16 From Dynamism to Inertia, Without Cuts! 91
- 3.17 Chizar Cultural Center ■ 95
- 3.18 Cube 97

3.19 Creation of Fluidity in Space 101
3.20 High, Light, Transparent. 109
3.21 Moving Surfaces. 115
3.22 Illogical Sincerity 119
3.23 The Color of Grey Mind. 123
3.24 My Scheme. 129
3.25 The Second Experience. 134
3.26 «Collage», Second Sequence 138
3.27 Large Scale «Collage» 142
3.28 Reinvestigation in Labyrinth of Mind. 148
3.29 Spatiography. 152
3.30 A Trip to Inside 160
3.31 Me and My Feeling, All Alone. 164
3.32 Scattered Ideas 168
3.33 Creation of Space 172
3.34 My Neglected Pictures 176
3.35 Two Case Studies. 180
References. 181

A Brief History of Collage

1

The word Collage is derived from the French colle or coller meaning to glue in visual arts. It is a work made by assembling various forms to create a new whole. The spectacular and brilliant form of this method was first introduced to oil painting in the early years of the twentieth century as an innovative art different from background. This term was then coined by Georges Braque and Pablo Picasso as a part of modern art. A work of "collage" is made of newspaper or magazines clippings, ribbons, painted parts or hand-made papers, parts taken from other artworks, photographs and suchlike glued onto a hard surface or a canvas. In an innovative move, Jonathan Talbot, inventor of the "collage" with multi-dimensional structure, omitted the glue from the process of making "collage".

A Collage Mixed Media

1.1 The Origins of This Method

Technique of collage was first used at the same time as paper was invented in China, around 200 BC. The use of collage, however, remained limited until the tenth century when the Japanese started applying papers glued together to write poems. The European started using collage in the thirteenth century (medieval). At this time, the gold leaves were used in the fifteenth and sixteenth Gothic cathedrals. Gems and other precious metals were used in images and religious symbols and also to hide arms.

Still Life with Chair Caning by Pablo Picasso (1912)

A. A. Adibi, *Collage: A Process in Architectural Design*, University of Tehran Science and Humanities Series, https://doi.org/10.1007/978-3-030-63795-8_1

In nineteenth century, this technique was used by hobbyists for memorabilia such as photo albums and books, e.g. the works by Christian Andersen and Carl Spitzweg.

1.2 Collage in Painting

In cubism, Pablo Picasso was the first to apply the collage technique in oil paintings. In 1912, he pasted a piece of oilcloth fabric with a design of a cane chair in his work called "still life with chair caning". Surrealist artists continued using this technique. Cubomonia is a type of collage which is created by cutting an image into square parts and then assembling them randomly and unconsciously. "Inimage" is a special style of surrealist collage which was invented by "Rene Paseson". But it might as well be called "decollage" in which some parts of an image are cut out so that another image can be illustrated. "Etrecissements" is the name that Richard Jonoz used to refer to collages which Marcel Marien first using the same technique presented.

He also introduced the "collage exploration" which consists of the "decollage" elements and was made of layers of painting images whose corners were glued very loosely and then some parts of the

Fruit Dish, Bottle and Violin by Pablo Picasso (1912)

Two collages by Le Corbusier in architecture

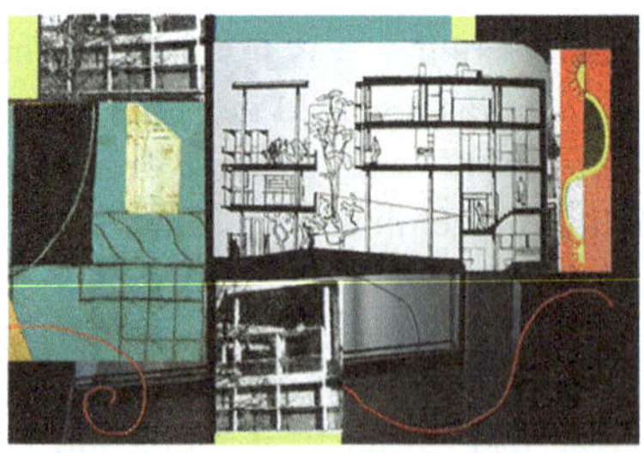

upper layer were torn off to show the lower layer, so a new type of "collage" came into existence. Penelope Rosemont invented some surrealist collage techniques termed "Prehensiilouette" and "Landscapade". "Collage" is often considered as an artistic form of the twentieth century, but in fact, it was never fully materialized. In surrealist forms like "Parallel Collage" a set of techniques are applied. "Collage Canvas" is another technique in which pieces of painted canvases are pasted on another canvas. In late 1970s, "John walker", a British artist, created some prominent works applying this style. This technique was also used in mixed media works of the American artist, "Jane Nezalk". "Lee krasner" was a highly self-critic person who repeatedly destroyed her own paintings by cutting them into small pieces and then rearranging them to create a new art.

1.3 Collage in Architecture

Even though Le Corbusier et al. have benefited from techniques similar to collage, "collage" as a theatrical concept came to attention only after "Collage City" by "Colin Rowe "and "Fred Koetter" in 1978. These authors did not support the visualized feeling of "collage" nor did they do anything to protect it from a conceptual collapse. But they applied the "collage" with its nonlinear orientation towards history and in the face of modernism uniformity as a tool to revitalize the designing process. It included the urban historical fabric and also helped the designers to find a way to improve their performance.

Colin Rowe is a member of an architectural group at Texas University with which he used this method in teaching. Another member of this group was Bernard Horsley, a Swiss architect, who continued his work to become a well-known professor in Zurich University. Although "collage" was more of a metaphor than an actual act to Colin Rowe, Horsley actively used "collage" in some parts of his designing process. He had a close relationship with Robert Slotzki, an artist from New York, and always discussed the "collage" and breaking in his workshop.

1.4 Wooden Collage

In the mid-1940s, thanks to Lois Nolson's 15 years of experience, "Wood collage", one of the recent types of "collage", started to develop. He created his statue-like "wood collages" using furniture pieces, boxes and wooden grids, railing, chair or ornamental margins.

In the catalog of the museum of modern arts, it has been written about Sky Cathedral by "Nevelson" that "it looks like a square airplane from the front, Sky Cathedral has the pictorial quality of a painting …" these pieces are seen as monolithic walls or rock pieces that sometimes can be looked at from different angles. Wood collages are much smaller and, like a painting, can be framed and hung. Wood plays the central role in these works;

Sky Cathedral by Nevelson (1958)

Wood Collage by Jane Frank (1964)

the wood is cut into slices and assembled piece by piece on a blank canvas (which can be often integrated with a painting) or a piece of wood.

These frames, paintings and wood-relief collages can provide the artist with the opportunity of taking advantage of the rules language and historical reverberation ruling over the image creation practice to hang on the wall. This technique can be also integrated into painting or other forms of art in a unified artwork.

In many cases, the natural wood such as rough woods and tree branches, processed lumps or even bark is used to make a "wood collage". You might ask if we can consider these works as actual collages. It is because the original forms of collages were mainly made by pieces of paper, pictures or things made by people that could sometimes embrace special cultural concepts. Since a chopped up chair or a broken ladder, which were used in Nevelson's work, can embrace the original and cultural defined contexts, they can also be considered as a potential element in collage. Probably the natural and unprocessed woods, like the ones found on forest floor, lack these contexts. Some woods are rough, ambiguous and unidentified, yet they can be found in a woodwork. Like a shiphull which has been so corroded and rusted by seawater that its past functional identity can no longer be maintained.

1.5 Collage in Music

The concept of "collage" has crossed the boundaries of visual arts. The technologic advances in voice recording have enabled the avant-garde artists to cut and join audio pieces. This move started in the mid-twentieth century, and during the 1990s, "musical collage" became prevalent in electronic pope music. The collages made of recorded pieces by George Martin can be mentioned as an example.

1.6 Digital Collage

Digital Collage, utilizing computer soft wares, integrates the incongruent visual elements to transform the visual outcomes of electronic media.

1.7 Literary Collage

Novel collages are created through selecting and assembling the subject matters, narrations and selected images from other works. "Discordianism" whose author introduces it as a literary collage belongs to this category. A literary collage can be also used to classify various visions and views about a subject.

1.8 Decoupage

Decoupage is often defined as a work of hand and includes placing a picture on a subject to be used in interior design. Decoupage causes the picture to be completely exposed or makes the subject and the picture seem perfectly matched. The process of the decoupage begins with a picture pasted on a subject, then more copies of the picture are added, to the extent that the main setting is entirely covered and profound connotation of the picture is exposed. The final picture is often protected by a coat of varnish or other similar substances.

1.9 Photomontage

Photomontage is a type of collage made of two or more photographs or some parts of photographs. Photomontage is a process through which some photographs are cut and glued to make a new composite image. Sometimes, the final image is photographed to create an interesting and unified product. Today, special computer soft wares are produced to assist the artists with photo editing. This technique is called "combination" by professional artists and "combination printing" for informal internet application. Victorian "combination printing" is another technique to integrate the photographs which contains the printing more than one negative on a single piece of printing paper (e.g. Rejlander 1857). "Front-projection and computer montage are named as other methods". Sometimes a combination of these collage techniques are applied by artists. A series of black and white "photomontage" by Romare Bearden (1912–1988) can be mentioned as a perfect example. He started his work by combing paper, painting and photograph on a board measuring 11 × 85 inches, and then using a hand roller and some emulsion paint he fixed the photograph, and as the final stage, he photographed and enlarged them.

The traditional method of physical combination of some photographs and photographing them and its uses in lithographic printing, prevalent in the fifth century, came to an end with the advent of digital editing. Today, editors produce the photographs of "layout" magazines digitally.

With the advent of computer soft wares such as "Adobe Photoshop", "Pixel Image Editor", and "Gimp" this work has become much easier. These soft wares accelerate the workflow and also produce more accurate results. Furthermore, by providing the possibility to correct the mistakes via "undo" option they have enabled the user to perform more effectively. However, some artists still face the dilemma of choosing between the traditional art and these soft wares with all their astonishing image editing and combining capabilities. The current trend is to create the images that combine the painting, theater, illustration and graphics in an integrated photographic whole.

Pearblossom Highway by David Hockney (1986)

Cloud Iron Designed by El Lissitzky (Photomontage)

The Secrets by Romare Bearden (1964)

2 Objectives and Steps to Make Collage Images

In this collection, we aim to steer the architects toward the challenge of self-analysis through introducing the collage technique. The challenge that familiarize them with the hidden aspects of their material and spiritual life and direct them to create the spaces that will be with them forever. In doing so, we should have a throwback to our hidden memory to discover the rich and life-giving spaces which are strongly intertwined with human entity. Now we intend to observe this technique closely and despite the intricacy embedded in each design, learn how this technique, like a replicating cell, forms many activities and creates an integrated whole and becomes a manifestation of the fundamental order rooted in our nature.

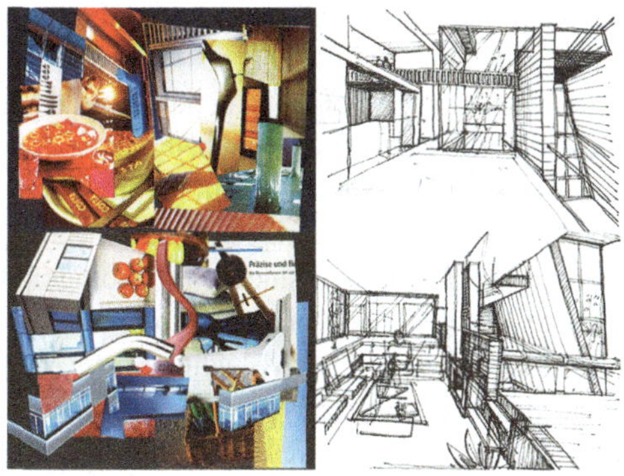

The challenge of materialistic and spiritual pints of designer, to create a space which is related to his/her characteristics

2.1 Advantages and Objectives of Collage

This technique pursues two objectives:

1. Boosting creativity
2. Building self-confidence and self-belief.

Agility, attention and attraction to mysteries and unknowns are the requirements to receive new subjects. It also requires a risk taking and exploring attitudes and of course, abandoning rigid and inflexible rules so that one can take creative actions when encountering new and unknown situations and based on his or her reasoning choose the best approach. In other words, one can train his or her mind through the results which are themselves the product of thinking. Conveying the spirit of science in training and education exceeds the increase of knowledge. It means cleansing the learners' souls from the diseases that might distance them from the truth and the spirit of truth seeking. The teacher's objective must be training the learners' intellectual power and vitalize the creativity power and independence of method in them (Motahari 1992: 54). In architecture training these two issues, creativity and independence in method, are considered as essential qualities required to be a successful architect. According to Amiral-Momenin Ali (a.s) in Nahjolbalaghe, there are two types of sciences: Heard Science, the science obtained from an outside source and Pleasant Science, the science derived from human nature, the science that is not taken from others and is the product of one's creativity (ibid., p. 19). Because of the flaws in

A. A. Adibi, *Collage: A Process in Architectural Design*, University of Tehran Science and Humanities Series, https://doi.org/10.1007/978-3-030-63795-8_2

architecture teaching system, and not the lack the talent, some architects are deprived of the pleasant science; the flaws that prevent them from fostering their talents. Therefore, undoubtedly, the main purpose of teaching all types of sciences, architecture in particular, should be to nurture the learners' mind. The teachers must impeccably fulfill their responsibility for nurturing students' minds and creativity and analysis power (ibid., p. 24). The purpose of making special images in collage is improving thinking abilities, boosting creativity and innovation. The images are the means to introduce tangible facts. There are various functional ways to achieve this objective but what matters is the possibility to convey the result. In this technique, visualizing a space without having some images of it is believed to be an impossibility, regardless of whether they are subjective and emotional or they are chosen beforehand. These photographs act as means of fostering the creativity, since the photographs used to make a collage are not final and unchangeable but subjective which have been emanated from designer's nature and subconscious. Collage is the expression of an architect's inner and unconscious thoughts which cannot be expressed through logical or conscious methods. It can be said that collage is the transference of the architects' ideas from inside to outside. The architects who cannot express what is going on in their minds or are even unaware of what they want and using logical conventional methods try to spark some ideas. Losing the original idea in the turmoil of chaotic and standard lines or expressing it so faintly that might be ignored even by the designer, thus taking refuge in standard designs can be mentioned as some results. But collage helps the architect to find on the outside what he or she was looking for on the inside by unconsciously generating some of designer's ideals, and it will be the reward of boosting creativity of inner memory. Collage does not cause creativity in designers, it mainly aids them to develop their internal abilities through introducing the right means. Of course, it does not mean that there are no other ways to turn the original idea into a design. It merely denotes the higher success rate compared to other methods.

This technique extends the boundaries of thinking and helps the designers to imagine beyond what they have in their minds and look beyond what they can see. Looking carefully at each phenomenon can result in new ideas. Looking closely at an image enables the designers to make an emotional connection with its components and simply understand the ideal feeling needed in a desired environment. The feeling experienced by understanding an invisible object in collage is as pleasant as feeling the real environment. We can easily shape collage images in our mind and even create a place where we can blissfully walk and open a window to our heart through which the self-created breeze will move our mental waves. Collage is the release of repetition tying designers' wings and preventing them from flying.

Collage is one of the creative ways to design which builds up designers' courage and because of its beauty, attraction and color variety helps them to stay energized. In reaching the goal, it creates a kind of ambiguity that can be considered as the secret of its success. The ambiguity that helps the designers to develop curiosity and improve their intellectual abilities.

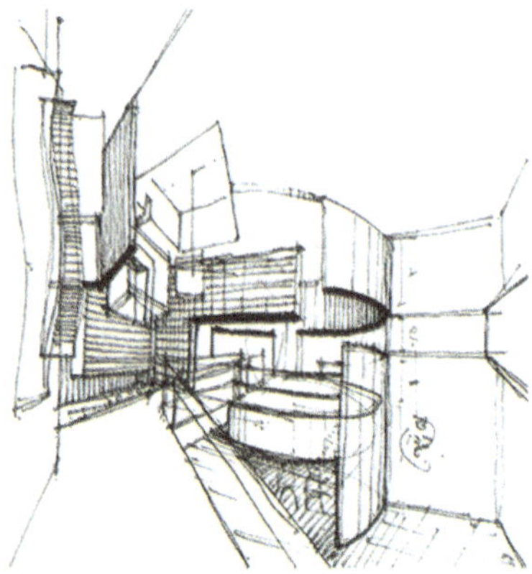

In these two images, the designer meticulously is exploring his hidden memory

Collage provides us with the opportunity to know ourselves better and realize the fact that we can also create impressive spaces for the first time indicating self-confidence, self-belief, potential talents, avoiding interminable arguments, and creating an extraordinary and conveyable design.

2.2 Tools and Materials and How to Use Them

1. Selected photographs by designer with regard to the desired subject
2. One sheet of A3 polystyrene
3. Some A3 black or white pieces of cardboard (preferably black)
4. Paper and scissors
5. Some pins
6. Some glue.

First, fix one piece of A3 black cardboard on an A3 polystyrene, then choose the appropriate photographs regarding the subject and pin it on the working board, and then reevaluating the chosen photographs repeat the process. It is worthy of attention that after pinning each new part and evaluating its composition in comparison to previous photographs and, if need be, changing the direction or location so as to reach the best composition, the working board is placed at a 3-m distance.

This tree-meter distance is emphasized because the purpose of making such photographs is not to reach an excellent combination of colors or contexture, but to reach ideal spaces concerning the design subject and this distance helps the designer to pay no attention to the details and focus on a whole different from a group of parts. For example, looking at the row of chairs or table at close range will reveal the fact, but when looked from a distance, it is associated with rhythm, recurrence or even a kind of flooring, and if the top picture is turned upside-down and looked again from that distance, it will be associated with fake ceiling.

After installing the photographs and making sure that we have the right combination, we glue the photographs on the desired place. Subsequently, we detach the black cardboard, which has now been turned into a colorful board, from polystyrene and get ready for the next collage.

When this photographis looked at from a 3-m distance, the row of tables can be associated with false ceiling and when it is turned upside-down, it can be associated with flooring

2.3 Design Method in Collaging

The design is not a grammatical and pre-fabricated trend, but is a product of mental and objective activity with a complex mechanism of creative action and rational action. It cannot be claimed that there is an absolute formula for design trend and this trend is explained in a single manner. As is evident, though designing is a process that is called "designing process", needs a sparkle to start. What a beautiful thing this sparkle is within the designer. From this point of view, collage in design process is a method for the manifestation inside the designer, which is explained in what follows.

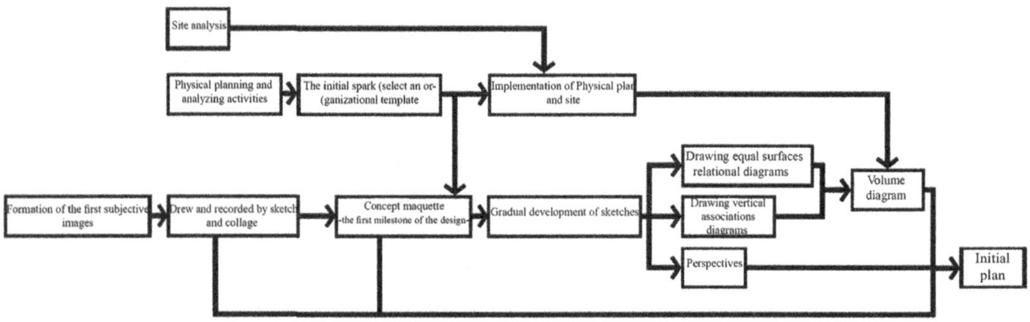

Before starting any collage, the design subject should be deeply thought out and all its aspects be considered and become familiar with its basics so as to view the designing subject, users, analyzing existing samples, etc. in a clear way. In the other hand, the individual should be fully prepared psychologically when selecting images and starting the collage in order to act freely. It is recommended that if the person is in inappropriate conditions and under psychological pressure, postpone creating images to later. Before starting collaging, a scenario about the subject or images mood, because these are as the continuation of fire flames which sparkles suddenly: just as when the poet does not know the poem form before being revealed to him. Collaging has also a poet-like nature: that is, it may be that the designer has no vision of what he is to design exclusive to him, but during collaging the images blossom and the seed of that vision gives fruits.

So one should not fear collaging process, but just as the designer works on the images, he should allow the images to work on the mind of the designer, give shape to it and lead it the way they wish. For example, maybe the result of the image expresses a kind of however, the type and intensity of the relationship with collage exercise is various depending on the character and inner belief of the individual treatment with design subject and has a modern or traditional mood, or may rely on separate concepts of architecture and has specified its prominent aspects of the work, carries the message of the plan or explains concepts such as rhythm, pause and motion, contrast, tranquility, balance, symmetry, etc. In any case, when this sparkle occurs it can be prolonged and give wings to it by selecting images reinforcing it.

This exercise is in fact a selective sketch. the only difference is that, in the sketch, the main tools of the designer is his thought who uses his inner self to express it so that he may manifest it as a work of art. However, in this method the eyes also helps the thought and sees the thing nurtured by the thought based on information and data of the vision. In fact, in this method selects the data harmonic and coincident with the subject among numerous data with regard to personal view and interpretation of the subject. Never data selection should be considered accidental. Maybe the evident difference of the obtained results is neglected by assuming that in this method or other common methods of designing the goal is to reach the final plan, but It should be noted that in sketch, the designer seeks help from his own unlimited mind but he faces problems because he cannot visually express the wide and unlimited range of his mind simultaneously. In this case, he seeks help from collage method since the creative mind of the designer has more tools in its disposal and can compare and contrast choices with design subject simultaneously with regard to its properties. In collage process, there are two different views regarding design subject:

The first being that a completely abstract and 2-dimensional treatment should be employed without investigating real criteria turning the idea into a practical pattern. Later, i.e. the stage of sketching the images, the mind is allowed to turn what is abstract and unreal into a pragmatic and real reality to provide more tools of a single plan in order to arrive at an objective reality near or away from what has been created.

Changing the abstract images to real usage

The other view is that ideal mental images of the concerned pragmatic environment be created through adherence to the realities of the pattern and through a rational thought and a 3-dimensional vision. The advantage of this view is that it reduces the designing mentality, because through using pragmatic tools and choices, the individual can convert his mentality into objective things by passing this stage; but its difference with the abstract view is the restriction of choices because of considering the scope of the work when exercising.

The images were chosen based on design and their perspective

When one of these methods is selected, the road of forming an ideal sphere starts. However, there is a fundamental question and it is "what is an ideal space?" the ideal space is a space in which humans achieve their wishes from corporeal and spiritual aspects and obtain the maximum level of tranquility: a space releasing humans from mental preoccupation. This is a pure and special space, which cannot be secured without considering inner realities and wishes. The pages of many magazines turn and many images are seen; favorite images are cut and separated from the magazine and are put in a file. Some images have a favorite color and light and others texture and material, and still others favorite forms, others only depict the lines of designer's mind and express his inner feeling. The designer is free to select whatever he likes. What he selects among hundreds of thousands images, shows his tastes, spirits, thoughts and character and not anyone else, and the very freedom restrict the arrangement of images to himself. Now, the designer reviews the culled images with an extraordinary accuracy and memorizes its mood, message and visual, technical, and graphical properties to use them when necessary and steps in the road of collaging. Then, he selects other collaging parts and through appropriate cuts and symphonious juxtaposing them creates new concepts in A3 cadre.

An inner force guides the designer in creating images without knowing it. The enterprise starts by a seemingly unreasonable cause to place special cutoffs in a special point with a special angle. Other images are selected with patience, accuracy and interest and are placed in other points. The designer distances from his images repeatedly, looks at it from a distance and from various angles, and dislocates the images repeatedly in relation to each other to achieve his favorite combination. Each time, one piece is added to the collage, from a distance of three meters looks at it to visually criticize the combination of it with other pieces, and the images are not fixed until the end, that is, arriving at the favorite combination. After completing three works and becoming skilled at selecting images and arranging them, he starts to look at and charting images; to see the images side by side is a kind of sport, the result of which is the discovery of favorite spaces, which are proportionate to special and spiritual spaces and moods manifested in a new form. The images created in this way, despite the efforts for creativity compared to previous images, again is in the same framework: a framework seemingly stabilized by the same force. In fact, in the route of collage the deciding element is that mentality and aesthetic sense formed and stabilized within the designer during years. The thing that makes this method attractive is to select the images freely without any restrictions and limitations and the freedom of imagination power to organize them, creating a deep bond between the designer and all elements of the image such as light, color, material, etc. This attraction encourages the designer to create more images, since each time more belief is put into the capacities of the images being started and consequently the capability of images multiplies in spatial imagination. After creating a few images, the set of images are put side by side and are evaluated. It is as though the whole feeling of the designer, favorite colors and its pleasant mood have manifested in the work which shows the ideals of the designer; the ideals that guide him to select clothes, decorations, resident, jewelry, etc. The collages of every person is in fact his hidden mind being illustrated. The mind, which is pure, and numerous ideas have not had the opportunity to show off, which the designer himself had not been capable of imaging it. Images are the articulating tongue of designer's beliefs. Apart from color, which is the individualized essence of images, something as a personal signature exists in them.

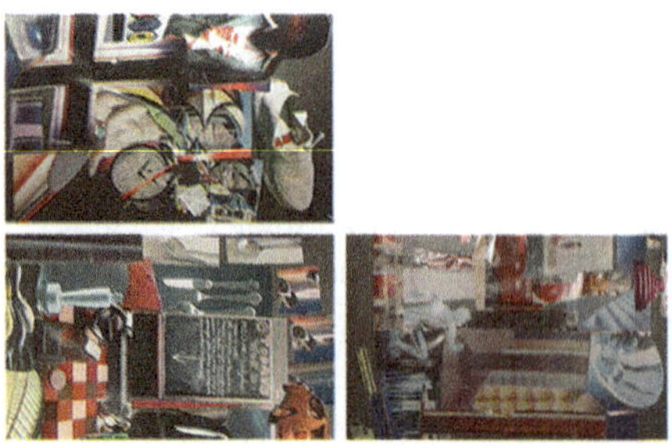

The manifestation of designer's ideals, favorite colors, material and mood

Tension in one point, protrusion in the other, sinking in the other, using hard or soft lines, liquidity or inertia, which they result in the person being satisfied of himself, all of these are the result of the relationship of designer with his inner world. The world of personal tastes and interests, recorded tendencies and ideas in the unconscious mind, etc. However, there are two important points in the arrangement of pieces: transparency and integration, which are inherent to collage and prevent the formation of boundaries between images, unless it is intentional. Images are put side by side smoothly and are integrated. They do not exercise pressure on the other, do not cross each other or suppress others' identity, but are placed next to each other to manifest the concerned concept and finally turn into a united set and its elements are juxtaposed in a way that none of them can be removed at all. In this way, the viewer's look moves from one image to the next one without being disrupted. The presence of these features in all parts of an image, helps the viewer to have a better perception of the concepts within the image. Thus, the images are formed in a context of transparency and integration; therefore, the image is perfect, and every time it is seen, it may convey a new message. Thus, images form in a background of transparency and integration; therefore, the image is complete and anytime you look at it, it may simply another message. Two-dimensional images that were intentionally selected with utmost relaxation, and were put alongside each other, now are used to create three-dimensional spaces and something beyond ordinary emerges in the sight of the designer and in this manner the design concept is fashioned. The most critical part of designing a collage is to convert architect's abstract thoughts into real spaces and volumes by observing proportions and scales. The important point is to be completely loyal to the process of designing the collage and to create volumes recognizable through combining images. The designer creates primary miniatures without the concern about examples and images, which are created later; thus, he creates new spaces and buildings and injects dynamism spirit into them. The reason is that when those ideal images take form, there appear a way to enter them into designing. One of the properties of this method is that you can see different persons with different personalities as well as different spaces and for everyone this can turn into an interesting and marvellous game to:

1. melt frosted minds.
2. Increasing imagination and creative power on mind.
3. Better view and a more critical view of spaces and reaching to the conclusion that "architecture is the combination of feelings and designer's thoughts".

The task of the architect is to understand the life of users with all their needs, wishes, and concerns in order to create spaces for increasing their life quality; every solutions it offers should be in the direction of that final goal, i.e. shaping a place ideal for human life; a place full of life spirit, dynamism, and freshness. In sum, architect's beliefs and ideals, which is a correct and good understanding manifested in the creation of place, and the more complete is this recognition; the more comprehensive is the result, which implies a recognition, which teaches the architect many lessons. The importance of this recognition is due to the undeniable influence of individual's characteristics, their inner essence and beliefs in the formation of their corporeal and spiritual life. Freshness and dynamism are two key properties of a favorite place and the more strongly these are achieved, the more inner forces are released and free him from the restrictions of acquired contra dictions and manifest in a unified dynamic totality. The more is the dynamism and freshness of a place, the will be more passion in the vessels of life, the precious result of which is to protect the inner self. Human being perceives his ideal space by knowing himself and analyzing it, which is the criteria upon which he judges the limit for the harmony of space with his needs and spirit. The space is perceived through the simultaneous activity of the body and soul and the thing that results in perceiving ideal spaces is the intuitive perceive or a supernatural power. The thing manifested in the mind wants to emerge in the external world. This concept is incarnated by the hand of the paper and is realized in a 3-dimensional shape in miniature space. The concept and incarnation are always simultaneous. The power and speed of the incarnation of this concept depends on designer's capacity that is different in person to person. Spatial creativity finds the opportunity to represent itself so that the individual does not fear and have self-confidence and good articulation. However, this does not happen unless with nurturing a good expression power. In a work of art, the necessities and necessaries are scientific and artistic capability, laws of physics world, and the spiritual world. When the architect discovers, analyzes, and corrects the spaces of his life maquette, can construct a new language and maquette in every new design and create an ideal structure for himself composed of constantly evolving structures and become "our" common language. Any factor challenged in the educational system for nurturing the creativity of humans is situated within the same creative analysis trend. Realization the inherent unity of all of their images and manifestations in creative efforts, which depends on an inner meaning. In this regard, no image is in seclusion. Any shape is an incarnation of a collage and any piece indicates an aspect of self. The only thing that is the result of inner feeling can have a spiritual meaning. The images created are an inevitable witness to the internal order of the designer rooted as a kind of architectural trend in all aspects of his life. Mutual relationships and manifestations represent all the arts and thus the paces represent themselves as ideal spaces of the designer and a single art.

In other words, we can say that a professional experience and a work of art depend on personal and creative skills. These properties rooted in the hidden memory of the artist, cannot be instructed. The thing that should be mentioned is that attaining the achievements of this method is dependent upon the harmonious application of all mental forces of the person not nurturing them separately; all of them should be nurtured simultaneously.

2.4 Color Painting in Collage

Different colors have different influences on the human's body and soul, and influence a large part of his visual understanding. Combining and merging colors unconsciously give dimension to images and create a space having 3-dimensions in addition to the fourth dimension, that is, the time. An immediate and unique experience deserved to be released from the 2-dimensional paper sheet and emerge in the stable form of architecture.

The thing that has happened is the manifestation of the consciousness on the paper through combining images and colors and poetic discovery of the space or the poetic space. The life hubbub was left behind red traffic light and twenty-first century man could not meet it but in dreams; this is our true "self". "A self that is original and free from all interests, pretentions, and routines. The ideal space of designer's mind revives in the creative and free combination of warm and cold colors, supplementary or contradictory, organized in an ideal balance."

The thing in this method attracts the attention of all experts, is the dissimilarity of even two images of two persons; because the whole thing has progressed freely from the very start, and the person selects his ideal color spectrum in relation to his unique psychological and personal characteristics and his understanding of the colors.

Color variety of the works of different students and color similarity in each work can be seen in these three collections

In this way, the designer penetrates into the hidden aspect of his "inner self". Ideal colors, ideal light, selected material, and contradictions and concepts; and in a contradiction and combination tangible to him, paints his ideal spaces and thus color is injected into the artificial environment.

Though the only meaning of collage is not a beautiful combination of colors, we can say that the combination of selected colors in collage derived within the designer is one of its achievements. The achievement that is sometimes in the form of a challenge-place contesting the designer to create an ideal space.

2.5 Collage Workshop Mood

There are various behaviors in a collage workshop. The work is started with a kind of ambiguity and curiosity. A special feeling emboldens those entering this atelier. Some look at the collage images unbelievably and with a smile and think that such a method for designing is impossible to exist from the very start. Others try to understand it. Thus, active students of the workshop with a special interest try to explain this method.

Getting started in the workshop is never been subject to the essential entry of the teachers to the workshop. In this way without the feeling of necessity to draw lines or to listen to the words of their teachers, each student can spend a long time on managing his/her works and offer various plans, and also the more student's knowledge of and relationship with their pictures, the greater their success will be.

The masters attend the works with a special humiliation. Masters listen to students carefully to teach them to transfer their work trend to others as best as possible. To listen carefully, to avoid mockery, and to encourage students, all have an effective role in establishing a mutual relationship. Then, efforts are made to find positive points the students have not spotted, insisting on which creates more trust. Then, negative points can be declared to the self-confident students without prejudice.

In fact, in this way, teachers are good listeners! They speak less or draw fewer lines. They put emphasis on the structure and functions of the space and if necessary provide required guidance. The 'piece seemed' to students is like an instrument based on which they design their perceptual spaces, and not the plan itself.

To achieve the images forgotten by the mind and their basics (although in that special field) requires a special accuracy. The collages continued to the end of designing process each time the students tries to discover a new space. Michel-Angel has said; "patterns and bodies exist and are composed of images full of space".

Sketching them and extracting personal perceived spaces of students, which even surprises themselves, is an exciting enterprise, but is unbelievable for others to some extent. The search for reaching 3-dimensional places is still going on. The images are broken into pieces and are prepared gradually. Each specific space and image that should be juxtaposed to create a single totality with regard to visual elements goes on. The images are hanged on workshop wall, refresh the work atmosphere, call the addressee to themselves like an open door, and create question in his mind and this curiosity causes the mind to think.

In the early session, the students are distressed because have little familiarity with this method but, in the other hand, experience confronting with the masters paving the way with caring encouragement. Students' images side by side and studying the differences in different areas and reaching personal beliefs, differences, and capacities the images of which have occupied a corner of their minds and a part of them displayed before the eyes of others and seeing oneself in the presence of the common people are discovered and perceived and this is the true life. The thing that an architect should grant humanity. All of these moods and experience add excitement to the monotonous and sometimes dreary process of designing.

The works in this workshop are conducted in a space of metaphor and suggestion and the masters give no inspirations, because in this method deductions are fully personal and unique (influenced by inner feelings, past and scientific knowledge of the individual). Every person perceives the space in a way that is not necessarily the same as others and the same thing creates variety in issues and generates many discussions in the workshop.

The variety of interpretations implies their rich visual and conceptual richness. Thus, it creates 9 live and dynamic workshops in which there are discussions, dialogs, and confronting constructive confrontations.

In this section, the thematic and schematic integration in three designers, works has been reviewed. These include synchrony. Overlapping, space interpretation, physics, physical proximity and foreground and background, regularity and irregularity, as seen in producing collage and designs of these architects. They use the artifact materials dependent on the space and time as a physical representation, to produce an interwoven form and texture. These designers produce their intended plan, regarding the safe conditions.

Robin Dripps proclaimed that: "As the junctures, seams, fissures and gaps in the figure are revealed, these become significant moments of discontinuity, small hooks grabbing onto the world beyond. Fragmentation and synthesis imbue meaning as a result of context, as the relationships between elements become more important than the objects themselves" (Shields 2014: 34).

By these designers, collage was valued as a tangible quality of space and form and highlighted through the process of producing collage. The partial transmission of transparency and material layers to motivate and engage feeling with work, evokes internal response with multi facet policies to interpret that architectural works present various and enriched methods in design process for the presence of collage.

1. **Le Corbusier**

> Observe with what tools man experiences architecture–thousands of successive perceptions creating his architectural sensation. His stroll, his moving around is what counts, the driving force behind the architectural happenings. The die is therefore not cast on a fixed point, one that is central, ideal, rotary and with circular vision all at the same time. (Le Corbusier describing the experience of Ronchamp, in Textes et dessins pour Ronchamp 1965)

Le Corbusier's collages were mostly paper ones which were a preface of the final work in the design process. He worked with painted newspaper, gauche and ink as seen in Picasso's paper collages. Le Corbusier used the collage in two scales, initial models and pictures with real scale. While evidence shows Le Corbusier's applying collages two dimension of works, the content of his collages include a regular ground in contrast fluid forms and lines and apparent transparency is used as a tool to signify the depth of composition.

Despite the powerful gestalt in Le Corbusier's architecture, synchrony and transparency are found in his architecture in the dialogs between form and ground. Quoted from Collin row, "The figure is simultaneously static and set in motion. There is the primary surface of attack, the frontal picture plane, and then, there is the convoluted and serpentine territory which lies behind" (The dynamic spaces taken from 2D collage combination in Le Corbusier's architecture are represented in crochet house, Austin villa and Marseille residential apartment.).

Le Corbusier, Taureau (1953) Tate, London/Art Resource, NY

In Le Corbusier's works, understanding interwoven spaces needs observer's movement (following the observation of a cubism collage in which observers movement demand synchronic viewpoints) the ambiguity in special hierarchy with space network situated in background, med ground and foreground lead to the repeated hesitation on the interpretation. Cubists considers the quality values of the single objects and applied it in the structure of collage which was followed by Le Corbusier. The art and architecture works of Le Corbusier were represented with apparent transparency theme, form ambiguity and composition which influenced the design, education and structure process of Hussley's collage and many others. Therefore, collage is not a piece of objects and result but is a process.

2. **Bernard Hussley**

> So collage is not only meant as an object, something made, a result, but what is perhaps far more interesting: a process. Moreover, that behind this way of doing something which as a result then leads to a collage, the collage could be meant as an attitude of mind. (Hoesli 1983)

Hussley points out that value in abstract painting and sculpture is the same as scaled object which is an immediate tangible method to help to assess the architectures space. For Hassle, collage acts as a tool to test the nature of professional exercises and educational approaches.

He describe his interest in the dialog between form and ground: full and empty are the same structurally in overlapping of interwoven spaces. The space around building and the space within the building have the same comprehension. This idea of dual space from continuity of ground forms with emphasizes on what the building and complimentary space shows from the hole comprehension as the characteristics of plan with the built sample, is the idea of modern space.

3. **Richard Meier**

Architecture is a long term process, starting from design and production to use. The design trend of a project takes 3–12 years while the collage construction ends up rapidly.

Meier chooses pieces in collage based on texture, colure, form, meaning, value, authorship

Bernard Hoesli, Untiteled (1966)

and connotation, showing that they are intuitively and internally planed.

The trend of producing collage is found in establishing some simple parameters. A general parameter in Meier's collage constitutes the main stream. Continuity and adaptability are seen in Meier's architecture and collage. Meier explains:

> In my work, I have come to rely more and more on the juxtoposition of that which one makes against that which one does not. This has to do with the play of architecture's inherent artificiality against an unpredictable and dynamidc context. I believe that this dynamic quality of natural and urban phenomena is very much a part of man's experience of architecture. (Shields 2014: 56)

Architectural phenomenology is an interpretation going beyond the simple analyses and creeps pluralistically into architectural works to realize the innovations. It is seeking to expand the sensual experiences and confront with work in the extended realms of senses. Mental images are tools which can promote this concept to turn architecture space into meaningful place; the hidden and apparent images representing meaning, identity and culture with this approach, the patch design method is introduced as multisense layering that moves beyond restriction and dogmatism to reveal the hidden sense in the work.

Year of construction	Architect	Analytical features effective in design	Collage
The three-fold combination of cow (cow tower) 1958	Le Corbusier	– The effect of regular ground of collage on apparent transparency of architectural work – Synchronic and transparency harmony form and ground in collages and architectural design – Perceiving intermingled space in the design such as collage requiring observer's movement	
Untitled (1976–1981)	Bernard Hussley	– Dialogue between form and ground in collage as the relationship of full and empty space in design (duality idea of space) – Using collage as a tangible tool in design – Ambiguity in architectural space resulting from multiple interpret from collage	
S 12, 1987	Richard Meyer	– Mutual effect of collage and architectural design – Juxtaposition of roles in architect influenced by element selection in collages – Simultaneous proceeding of continue and conformity features in design and collage construction.	

The architects apply the collage drawings as a tool to introduce geometrical order. Attraction defined by geometrical order in collage and the conceptual profundity represented by Miss van de Rohe, Daniel Libskind and Rem Kolhass layered shapes are based on structuralists, technics. These design-collages introduce 3D architectural compositions witch interweave the simple geometrical forms to highlight the materiality and spatial depth.

1. **Mies van der Rohe**

> The long path through function to creative work has only a single goal: to create order out of the desperate confusion of our time. (Honey 1986: 37)

Organizing the unrelated pieces in producing collage is in fact a relationship between collection of Mies's individual collages and the specific design trend showing his calling for order in chaos world. Mies used collage in the initial combination of design studies and perspective representation. Collage provides a means of interrogating materiality in the design process.

According to Mies, the dialogue between architectural logic, Scaling figures and surrounding landscape can even be seen in the first design sketches.

Drexler believes that in contrast to Constructivist and De Stijl artists, Mies used perspective with an abstraction which makes the ambiguity of interpretation of spatial features and materials of the proposed projects possible. He used collage structure to reveal the quality of materials and space of the proposed space along with a dialogue between architecture and natural context (Drexler 1960).

2. **Daniel Libskind**

"Collage challenges the notion of architecture as a synthetic construct—a mixture of various creative disciplines. The idea of achieving such a synthesis by a hollow manipulation of heterogeneous elements can result only in "synthetic flaws"... The analytic approach, devoid of multiple means, limited to the existing materials, determined by historical realities, can fashion a new framework for concepts and ideas by releasing us from the determination in both syntactic and semantic domains. It is a kind of poverty that is being proposed. A poverty of matter to be compensated by the development of the conceptual ideal. The use of garbage reveals a possible method which can penetrate beyond the surface appearance of meaning and form, and strike at the very depth of conditioned relations; generating through a new grammar radically different ways of perceiving, thinking and ultimately perhaps acting" (Daniel Libskind, in John Hejduk's Education of an Architect: The cooper Union School of art and Architecture, 1964–1971).

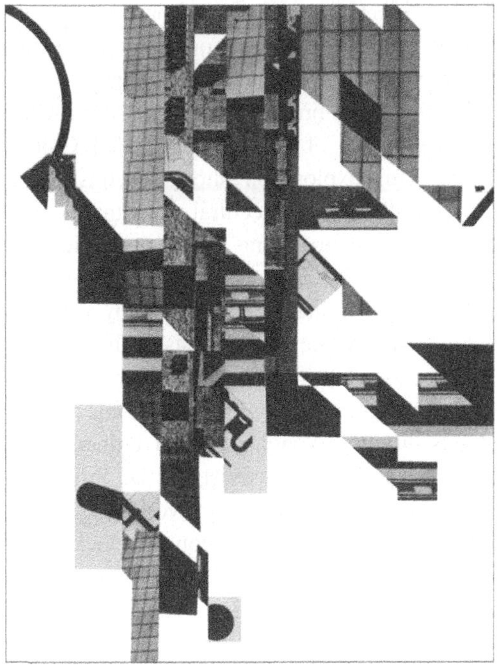

Daniel Libskind, Collage Architecture Structure (1970)

Libskind represented a method to change collage to three-dimensional form. His method defines experience of both construction and deconstruction of geometric order in architectural design which he examined it in collage.

He worked in project in Cooper Union which caused a series of explorations called Collage Rebus. This exploration shows deconstruction and separation of architectural plans that act as a main factor of analyzing of axonometric form and space.

Libskind's collage drawings employ a shallow pictorial space like the Cubists which express depth by displacement and fracturing of form. Then these compositions show tectonic investigations through axonometric that influenced by the projects of Constructivist and De Stijl architects. In Collage Rebus, overlaid grids—a orthogonal grid and a 30° counter-clockwise—provides framework to arrange complex geometric elements of designs. This geometry is resulted to three-dimensional form through axonometric designs.

Study of geometry and formal connections which are produced through abstraction of collage drawings, represents series of spatial interpretations that is examined through axonometric perspective. Libskind deals with apparent objectivity of axonometric and through creating plans and designs that make possible for multiple readings according to the ambiguity of spatial referents. This is obvious in Collage Rebus 3 which angled elements and rotation of plan is greatly reductive and dynamic spatial configuration is formed through orthogonal displacement.

Text—for Libskind—provides metaphorical purpose. He admit language as a tool for correlation with architecture. Selecting and ordering the elements, whether words or architectural elements, represents concept with the ability for multiple interpretation. Vesely explains: "The visual reality of Daniel Libskind's drawing is situated very far beyond the reality of cubism, constructivism, or collage, close to the horizon where most of the non-figurative movements of this century fought their last battles and where our imagination is permanently challenged by inner possibilities of abstraction" (Vesely 1981).

The created works of Libskind are kinds of evidence of applying line as a tool to arrange elements and unifying the form. The legacy of Libskind's academic pursuits within the pedagogical rubric offered by Cooper Union in the late 1960s is embodied by his built work of the

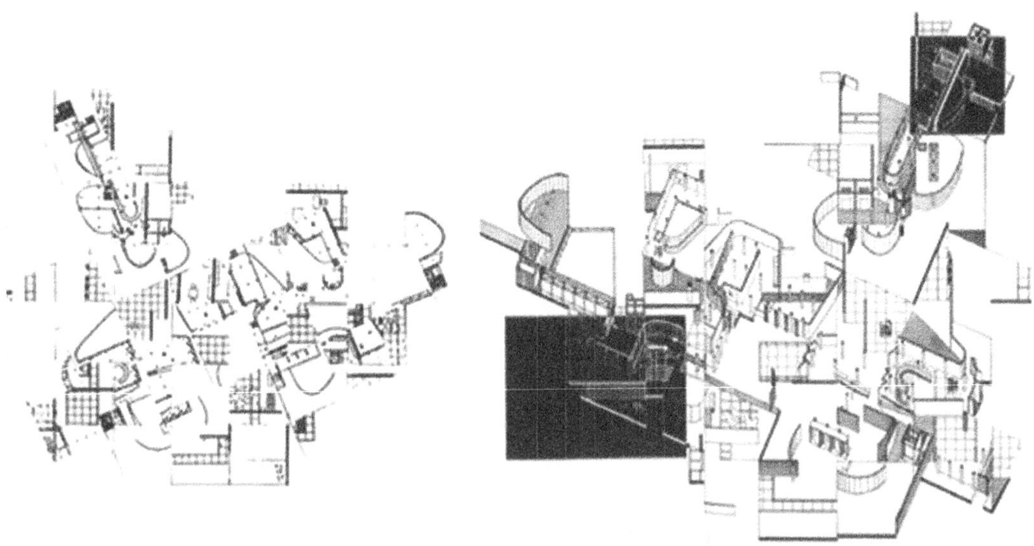

Daniel Libskind, Collage Rebus (1970), collage and axonometric

past decade. In his construction, he has physically and materially iterated architectural futures ordered and overlaid with history and narrative. We can understand the following strategy for collage—drawing as a preclude to his built work. Libskind wrote in "End Space": "An architectural drawing is as much a prospective unfolding of future possibilities as it is a recovery of a particular history to whose intention it testifies and whose limits it always challenges" (Libskind 1990).

Koolhaas's interest is revealed in a reciprocal program linking the deferent applications of a collage in which he searches for combination of unrelated elements.

According to Koolhaas "by not stacking activities, but by positioning each programmatic particle as part of a dense mosaic, our building contains the urban condition itself" (Koolhaas 2004).

Through this, a collage can be act as a tile or a combination of unrelated elements of a plan. The composition attracts plan layers and circulation

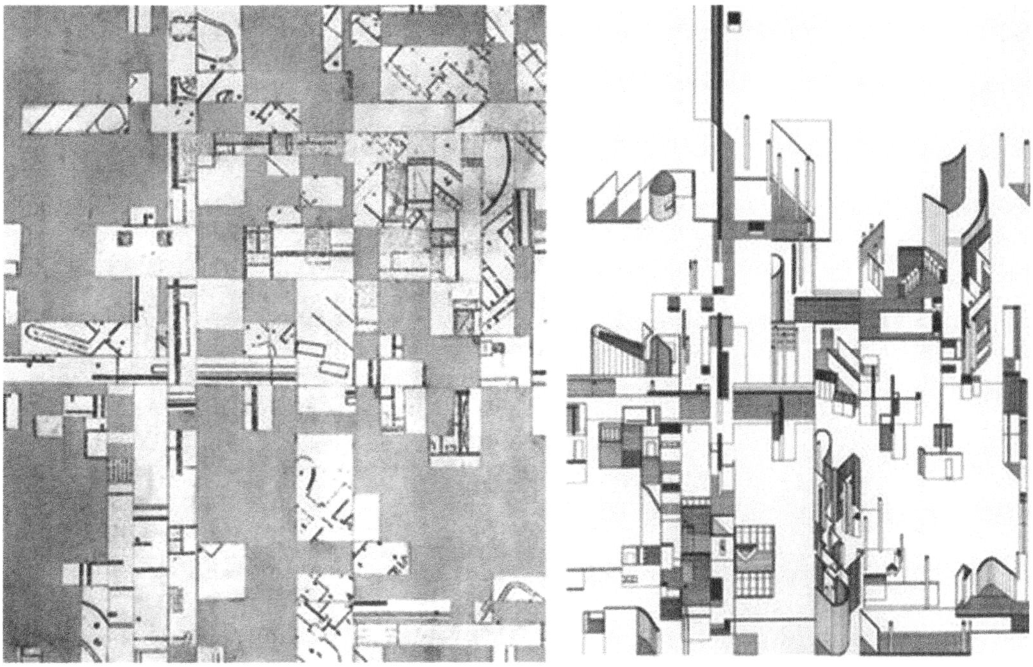

Daniel Libskind, Collage Rebus 3(1970), collage and axonometric

3. **Rem Koolhaas**

> If there is a method in this work, it is a method of systematic idealization; a systematic overestimation of the extant, a bombardment of speculation that invests even the most mediocre aspects with retroactive conceptional and ideological charge. (Koolhaas 1985)

to serve in different scales to develop as a city or internalize the city ground. The event space and narration as a collage demands interest and attention. In Koolhaas's work, time plays a role in narration and fluid strategy revealing a link between his remarkable work in AA and later architectural projects.

Year of construction/name of artwork	Architect	Analytical features effective in design	Collage
Proposed design for concert hall of William and Meier company	Ludwig Mies van der Rohe	– Organizing unrelated pieces in constructing collage represents thought style in design – Using collage to represent final perspective of project – Using collage to reveal material quality	
Pictorial architectural collage 1970	Daniel Libeskind	– Deconstructing geometric order and constructing pictorial architectural collage to analyze architectural design – Using ergonometric perspectives to analyze architecture design – Using line as organizing tool in assembling and integrating form	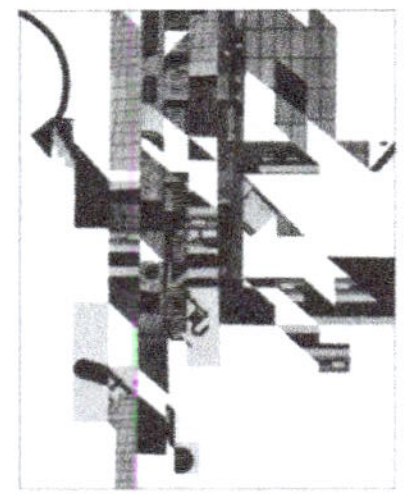
Mass migration of prisoners 1972	Rem Koolhaas	– Linking different application to different lagers of collage – Using time as the narrator of present design and future designs – Fluid strategy of narration with geometric order in collage and design	

According to Cohen, "Rem Koolhaas measures and slices the body of architectural history with his retroactive scissors," a process of decontextualization and transformation (Cohen, The rational Rebel). He applies a formal method of superimposing sections of geometric order to make and develop a space for human being activity within the narrative construct of social order.

3 Two Cases of Experienced Examples

3.1 ■ Architect's Housing Design (The Second Home)

Introduction

Human being and his living space are of a complex essence. The issues and elements to be introduced and analyzed when designing are numerous and is of different nature and properties.

In architectural design 2, in addition to a residential unit, there are other important educational goals such as nurturing creativity capability, self-confidence and self-esteem of students. Dealing with new issues requires speed and accuracy in one hand, and unknown phenomena in the other. The conscious part is the mind and spirit of humans. Risk-taking, searching and releasing from everyday contracts, which have gradually become customs in designing process, are the requirements of receiving the creative thinking in designing.

Goals of lesson

Despite the final goal of this exercise, which is that students acquire necessary skills for designing a residential unit and shaping it and creating the possibility of living there through a systematic and process according to the climate, culture, and needs of those living in that space, the most important goals of instructing architectural pattern 2 include awaking latent talents.

The process and stages of the work

The work is conducted in three stages:

1. Though release stage: in the first three weeks, students' talents are released and a belief is created in them in the first exercise through constructing and designing ideal and virtual spaces by a complete abstract and emotional expression through instructing and employing the collage method.
2. The stage of transferring it to others: the images are placed in a 3 m distance and are recognized and considered with a sharp look and ideal spaces are illustrated and offered through sketching and creating maquettes in a completely loyal manner.
3. Planning and generating meaning: in this stage in which students should pay attention to the site for designing a residential unit, in the first place students should offered their detailed designing plan to one of the masters based on theoretical basics and meaning of designing a residential unit, and when confirmed, they should consider the conceptual designing of unit volume based on the lessons learned from the plan background and other maquettes of residential spaces on the site.
4. Designing: after understanding the general meaning of a residential unit, its function design is carried out based on the plan and diagram of inter-spatial relationships in the form of plans, facades, cuts, and plan site. In this stage, arriving at favorite spaces in the exercised residential unit is possible. In this process, in addition to replying to the demands and needs of the architect family,

A. A. Adibi, *Collage: A Process in Architectural Design*, University of Tehran Science and Humanities Series,
https://doi.org/10.1007/978-3-030-63795-8_3

the students should consider creating and using the material, equipment, and installations for a reliable and relaxing life in the residential unit. It is necessary to offer the appropriate arrangement of internal spaces regarding the use of colors, lights, and contexture in order to create a pleasant space in the residential unit.

The program and plan properties:
The architect can think of this residential unit as if he should use this residential space in the future. He should consider a small family that would have a bigger family including two or three children (daughters and sons) nurtured in this residential unit. And even with the presence of a grandmother and grandfather, they live in an ideal residential space. The main spaces of this residential unit are as follows:

A. Residential space: the spaces designed by architect's family and himself based on their needs, which includes the entrance, living room, kitchen, bedroom and relaxing room, parents' bedroom, guests' bedroom, and toilets.
B. Workspace: the spaces predicted by those coming to architect's work space, which includes architect's office, small designing workshop, and peripheral services and parking(s).
C. Common spaces: the paces both mentioned spaces could use under certain circumstances, such as spacing for children and the yard, main entrance to the house, etc.

Site Plan
It is situated in a mountainous, cold region. Land slope has been assumed to be 15% and there are beautiful natural scenes. The selected site ends in a mountain, in the southern part and has also created the possibility to merge the site with hill nature. The estimated dimensions of the land is 40 * 25 m^2.

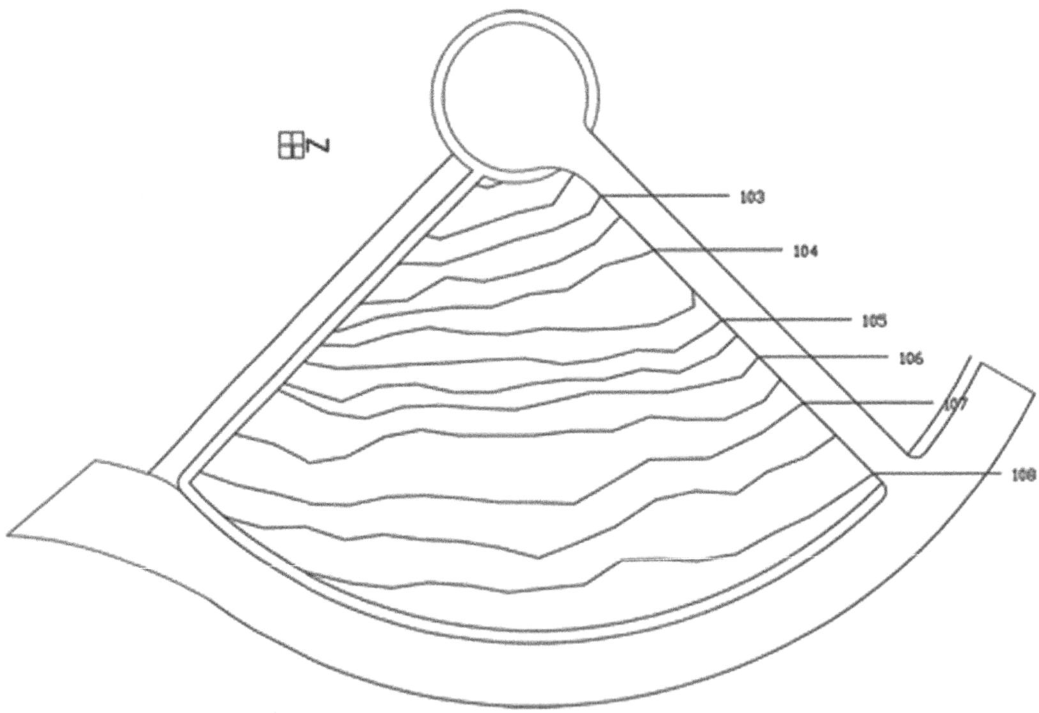

3.2 Axes Fracture

Low:

Space for studying

Next page, Right:

Collages

Next page, Left:

Sketches

Design of the architect's house

Student	Babak Shariati
Masters	Dr Ali Asghar Adibi
	Dr Ahmad Ali Farzin

There are different solutions for any problem; one is more rapid, the other more reliable, and the other one more creative. The problem can be solved more easily when we know different solutions.

In plan 2, the architect which designed architect's house, in order to arrive at the answer we assumed the problem has been solved. So that we wanted to design interior spaces and then develop it to arrive at the final complex.

This process was selected because interior spaces are the main property of a residential unit. The interior space is more important than the prominence of a residential unit as an urban element.

The effective factors in interior spaces include lights, colors, material and the combination of these elements.

The first process that stroke my mind was that a volumetric concept be sketched, the relations be solved, and spaces be designed in the plan and section. Finally, lights, colors, and material should be selected. But in this plan that the problem was assumed to be solved, at first lights, colors, and material were selected and then interior spaces were imagined and plan and volume were developed.

Thus, the collage subject was put forward.

The first thing to do was to prepare the clippings and pieces of magazines. The clippings were from pictures reflecting my favorite colors, lights, and gender. After preparing the clippings and arranging them on the table, constructed my collage. This collage was completely abstract and light and color combination was among the important things I controlled after distancing from the collage.

After creating one or two warm up collages, I assembled two main collages. These two pictures were more consistent with my mood compared to early pictures and this was the result of the experience in selecting and cutting the pieces in the early collages. The next stage was to sketch the collages. The details were less observable when sketching from a long distance. The interesting thing was that I drew a space I liked, which was due to the primary feeling of combining collages or forcibly extracting a space I liked. The sketching space meant with collages, because volume was in the sketch and color, light, and material were in collages.

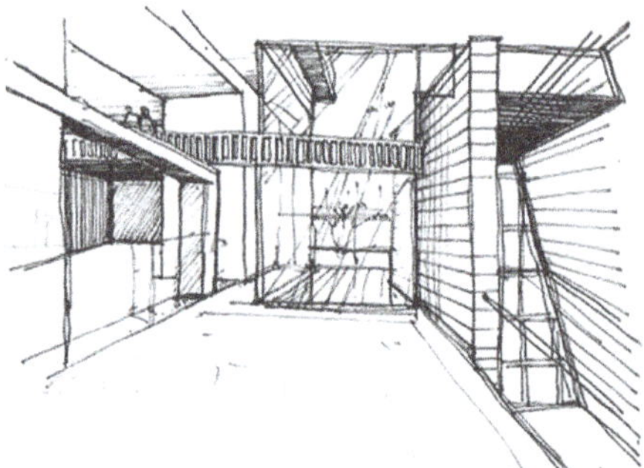

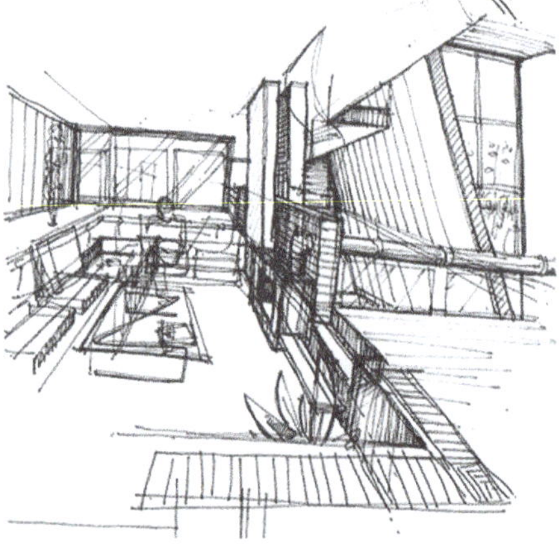

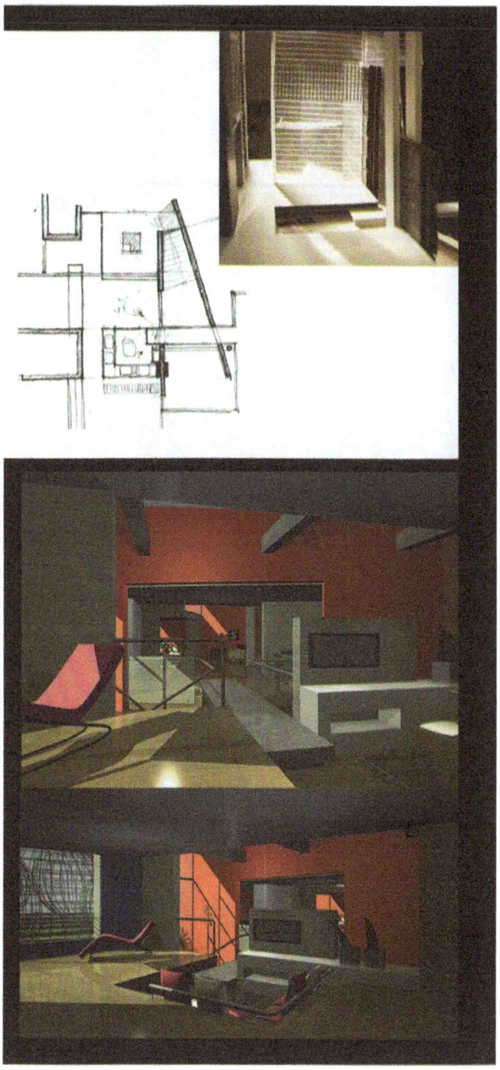

Sketches were completed and I drew spatial plan composed of the combination of two sketches. Finally, I creates interior maquette according to the plan and collages while trying to use the same color and light I had seen in the collage.

There is living room space in the interior maquette obtained from the collage, and this was the first sparkle of designing architect's house.

Here, explaining the house completed, the spaces were defined and the site was determined. I started to analyze spaces, their relations as well as site analysis and locating the space in it. Based on the locating diagram I determined interior maquette space in the site and extended periphery spaces from within. I would imagine the space and constantly conducted interior sketch so that spaces better take form.

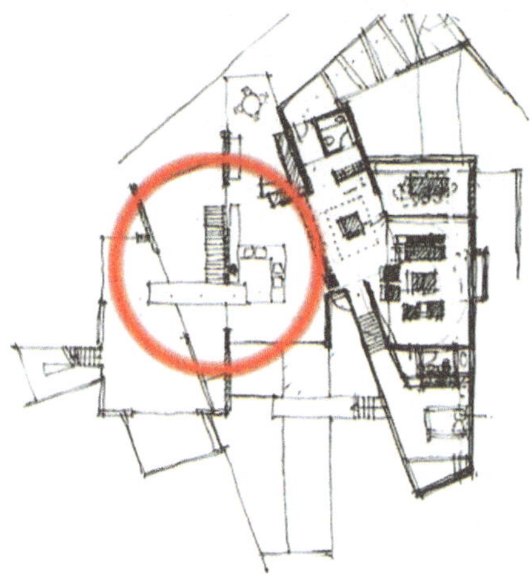

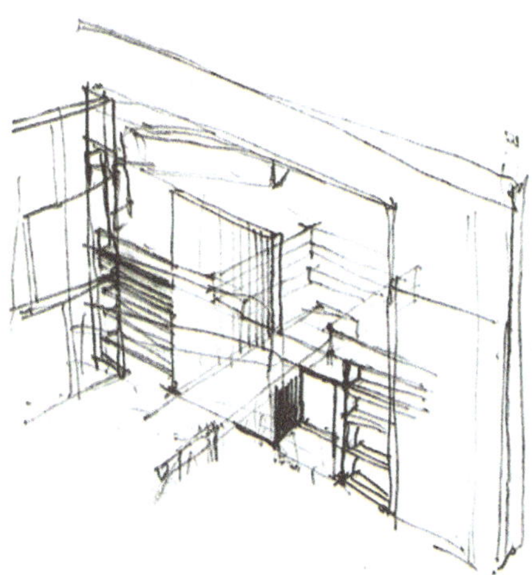

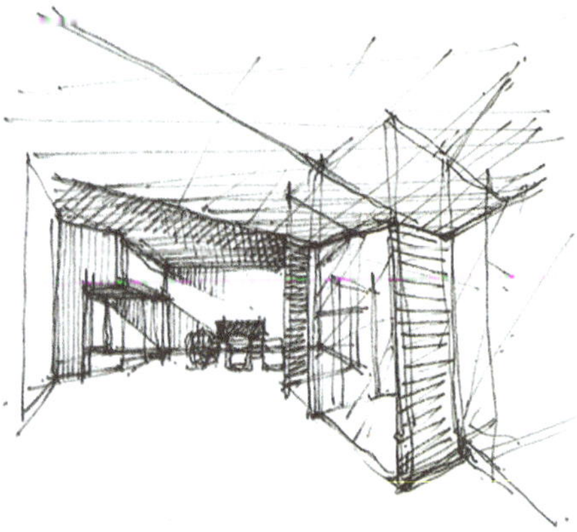

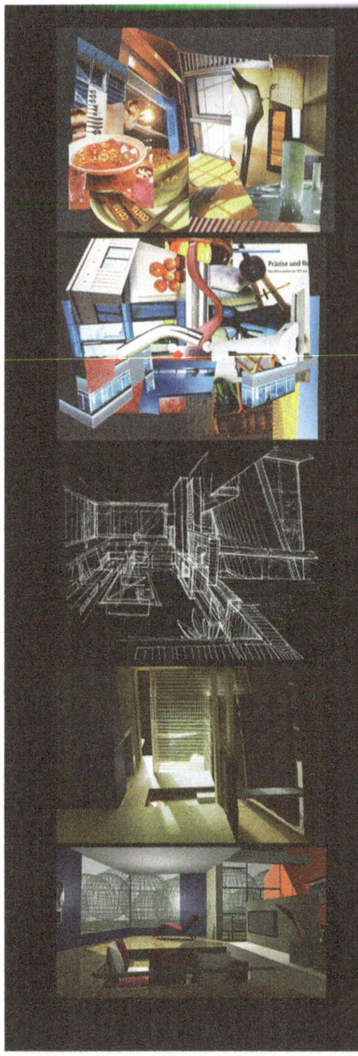

Upper left:
sketch and 3 dimensions of the concordance of final space with primary sketch

Left, top to down:
primary collage, second collage, sketch, interior model of sketch, Final 3 dimensional, front page, left

Top to down:
interior maquette, plan, 3 dimensional, front page

Right:
spatial locating obtained from the collage and extending the spaces the sketch for interior design

The idea I got from plan sketch, was the fracture in axes. Their repetition and connection helped the integration of spaces. The spaces were extended based on the diagrams of the relationships and locating.

There is a cold space in one of the collages, without any light, with concrete material and cold and low diversity colors, while in other collages the same colors have heated under the light of the supposed sun. Thus, the same concrete and sun of the collage warm space is obtained for the living room space.

In the same manner, the collage turned into a 3 dimensional volume by the process of sketch, plan, interior maquette, analysis, locating, and extension. The interesting thing is that only one of these spaces form directly from the collage and the rest followed the same system.

The motion of the crust helped to define the interior space in the concave section and the exterior volume in the convex part. Its red color coordinated the interior and exterior.

Spaces are classified into three categories: 1. Public ad workspaces, 2. Semi-public and service spaces, 3. Private spaces. Empty space distinguish group 2 and group three spaces.

The entrance was designed by considering reaching possibilities. After the entrance space, the divided space formed between public and semipublic space. Guests' room can be reached from the hall and, in emergency cases, from the northern terrace there is a small salon in the southern direction of the living room dedicated to work and non-intimate guests that can be accessed from the atelier and is separated from the living room by a Paravan like wall. Dining room is away from the living room so that the sound of arranging the table does not disturb the guests.

Private spaces relate to the exterior spaces through northern terrace, and in the interior space, it relates to sauna, the yard and atelier.

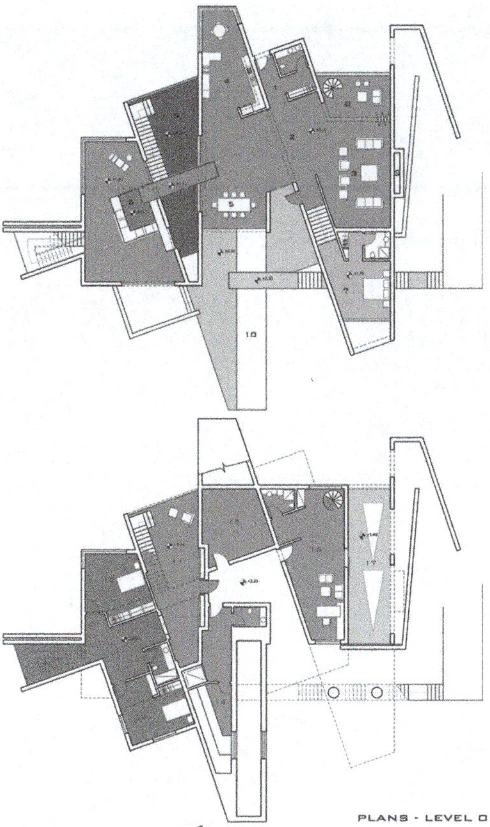

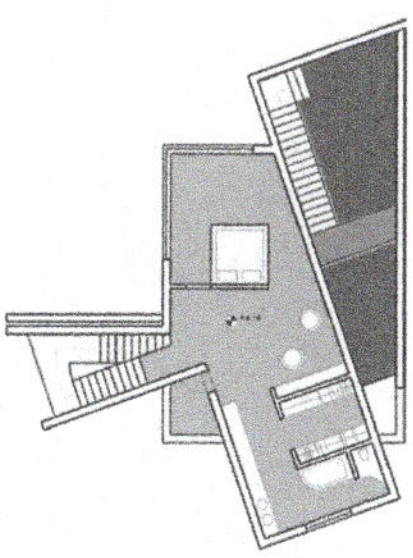

PLANS - LEVEL 0 : 1.ENTERANCE 2.HALL 3.LIVING ROOM 4.KITCHEN 5.DINNING 6. TV ROOM 7.GUEST ROOM 8.PETIT SALON 9.VOID 10.POOL LEVEL -1 : 11.STUDY ROOM 12.BEDROOM 13.CHILDREN'S PLAYGROUND 14.SAUNA 15.VENTILATION ROOM 16.STUDIO 17.GARAGE LEVEL +1 : MASTER BEDROOM SCALE 1/100

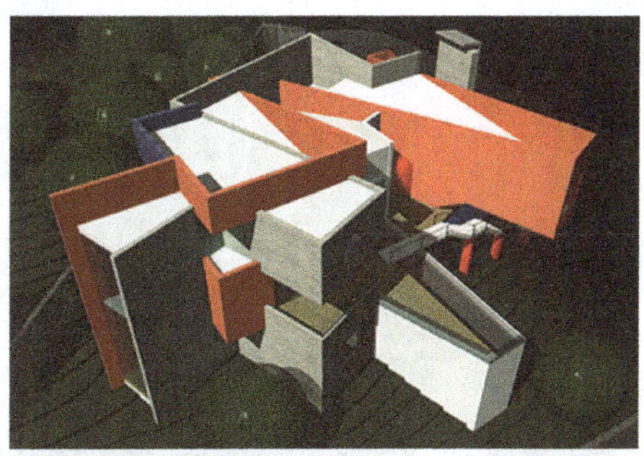

View of design

3.3 "Collage", a Novel Method in a Novel World

Design of the architect's house
Student Yasaman Haji Esmaeeli
Masters Dr Ali Asghar Adibi
Dr Ahmad Ali Farzin

Collage belongs to another world. I stand on the boundary of two worlds and create a supernatural space using material world tools; a space having eternal light and pure colors, a unique and unrepeatable space that only derives from my mind, not anyone else.

The stages to conduct the work
Firstly, we should more familiarize ourselves with the collage meaning. Any person draws completely different images and, more interestingly, no image was similar to the main collage because collage is like a window to the inner world. There we face infinite spaces hidden in our unconscious.

I draw the sketch with half-open eyes and look at the collages in a long distance; some lines highlight and the lines protrude from the page that are beautifully combined.

The motion of these lines and the continuity or fracture of spaces in each collage reminds me of a special space. However, in this stage the sketches belong to an imaginary world and have a long way ahead to become real.

In the next stage, I collect the images that attract me and juxtapose the colors and light calling me to themselves to create the space that should construct a house, a distinguished house for an architect.

With the supposition that each picture is part of my hidden dreams and are complementary parts of a puzzle, I put pictures side by side. All the things form beyond the boundaries of the worlds and lines freely join each other.

The light give dimension to the space and colors give life to pictures and therefor a space of my unconscious is created. Forms and shapes are completely in harmony and lead the mind into perceiving a new space.

In the next stage, I created a plan according to each collage indicative of the main space of that collage. This plan completed using colors and showing the gender and light.

These sketches were of considerable help to me. At first I did not know what is to be drawn, but as soon as the first line was drawn, other lines were automatically drawn on the paper by looking at the collage.

Though the collage itself shows the spaces hidden in it, sketches made these space more visible.

View of design

Eastern façade of the building

Installing the volume in the site

The first collage

Sketch of the first collage

General volume

The first collage

The first collage I created was related to work space; a room full of books, a live space with happy colors. Volumes turn and slide on each other, lines cuts create various levels, vertical lines break these pages into pieces.

A thoughtful, meandrous, and dynamic space. The light is broadcasted from every corner to the space. Apart from the small lights broadcasting a mild light to the space, it is the most important color in collage center coming out of the land core, a point the eyes are fixed upon. Flame dance inside the ground, fire reflection on the space. As if, things are thrown in a halo of light.

The red color in the center of the collage attracts me and creates an excitement mixed with reliability. Color change in the grey background causes bottom break and various levels.

There are no horizontal or perpendicular lines. I the collage, but any group of lines converged toward a special point, as though there is an extravagant perspective of the space.

In part of a high window, a natural mild light shines into the room, where I after a hard work the peace falls upon me by looking at the beautiful azure sky.

The second collage

The central space of the house was drawn according to this collage. In this collage, all pictures are put side by side in a completely abstract manner. The combination of red and blue colors is evident in the collage. As if the red pyramids and sheets hang smoothly between sky and earth. Time is broken in this space and flow inside these lines. It seems things revolve around a central point in the depth of these pictures. The space is illuminated with the lights extending to the central point. Vertical distances become dark in the sheet and consequently the space seems to be deeper. An illuminated flame in the center shines light into the space.

In the sketch extracted from this collage, the light levels seem to be nearer and dark levels are behind them. Lines and sheets gradually rise or join each other. The sudden fracture in these lines creates a new category.

The third collage

This collage, created for children's sleep and playing, is in fact the space of my childish happiness; a place for playing and laughing, full of light and full of happy colors. The ground is soft and again all lines go to one direction. A glass wall in the floor illuminates the space. The same fire flame in other collages shines in the center, but here it seems to be smaller and more intimate. There is a safe space for sleeping in the right corner of the collage and a blue and beautiful scene formed by green trees can be seen from behind the window glass.

The second collage

The related sketch

Volumetric maquette

The third collage

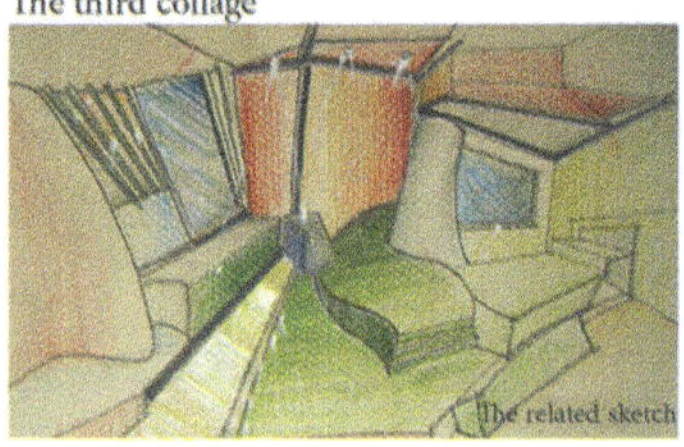
The related sketch

The related sketches The fourth collage

The volumetric maquette of workspace The collage related to workspace

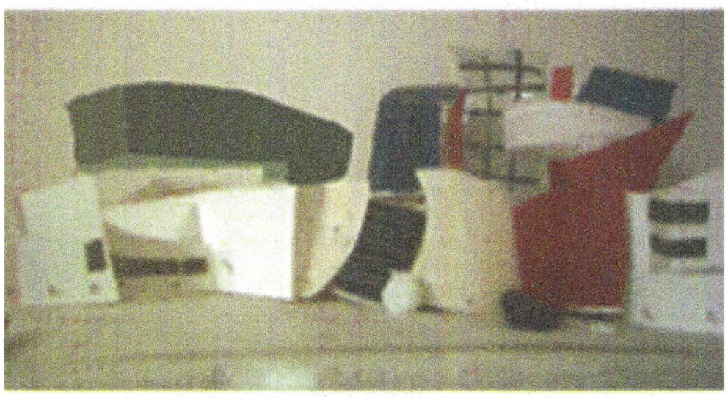

Collage and volumetric maquette related to the central space of the house

The fourth collage

This collage was especially created as an external space and for defining the entrance of the house. A luminous lightning breaks the cold metallic space and rises and opens a way beyond the metallic surfaces. In the end of this road, we enter the house entrance by passing illuminated roads.

Volumetric maquette

In the next stage, the plan progresses in a volume manner. The final volume is created after drawing the plan of each maquette and applying the necessary changes with regard to the form and function.

Workspace

All the properties of the main collage were preserved in the maquette, though an abstract space was created. However, when changing it and identifying the function, a rational space emerged. The maquette constitutes two parts of work office and the workshop, the workshop being situated above the work office. The entrance door is beside a big window and a mild light shines into the workshop through glass grooves.

Central space of the house

Like the related collage, also this maquette is completely abstract and created a different space. The broken clock in the collage defines the stairwell going up smoothly and ends in a big vertical window in upstairs. Every elements are fixed in their place but a dynamic space has been created by changing sheets height and twisting.

Live and spicy colors in the maquette indicate an intense and unending activity flowing inside the house.

In the next stage, after summing up the examinations about dwelling and related needs and identifying necessary spaces, other maquettes are created based on the primary one and gradually the function the sections of the space is determined and the main volume is completed by juxtaposing interior spaces.

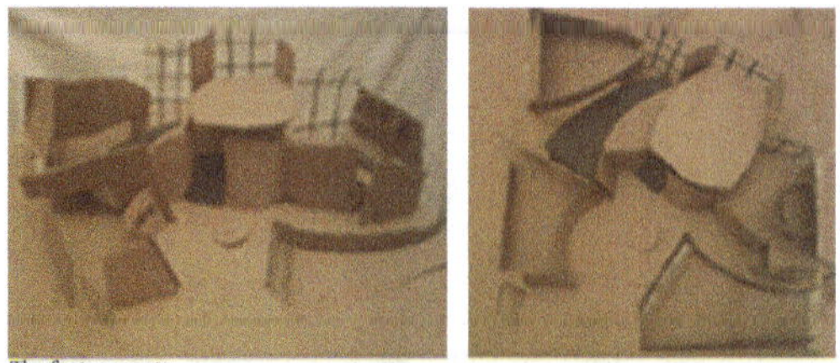

The first maquette

Image of interior

By conducting the required research about the intended site in Lahijan City, northern Iran, with 15% slope, I constructed other maquettes on the site and through creating necessary surface differences; the volume completely corresponded on the site.

The volume is created by two intersecting ovals in the spot where the two squares meet, so that the intersection points of the ovals correspond the sides of the squares.

One of these squares has twisted toward the other with regard to the directions of main lines of the site. A translucent coating has covered the first half of the second store and has created a spiral space protruding from the northern part of the volume.

This spiral space is heated by solar energy, and can supply a large portion of the energy of the house. By changing curved glass panels covering this space, when needed a flow of fresh air enters the house and ventilates the air naturally.

West view

South view

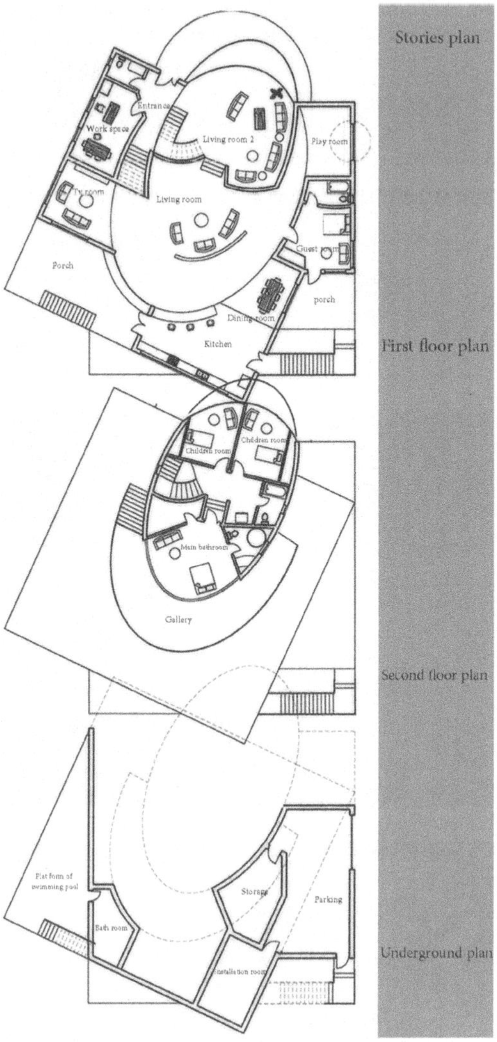

The spiral volume, which exists in the northern part of the building, has a very beautiful view toward the periphery scene.

This space as the gallery and library is closely related to workspace and sometimes is used as atelier and a place for thinking and attaining mysteries and new ideas.

Moreover, a small living room has been defined in this space, which is a gathering place for family members to enjoy the beautiful scenes and the pleasant light of the sun.

Guests' sleeping space is in the ground floor, that of the children is located in the first floor, and each floor has its own service.

The main sleeping space has been defined in the upper part of the volume overlooking other spaces.

Two independent entrances, one from the main pedestrian path (the main entrance) and the other from the parking (subsidiary entrance) facilitate the availability of the house.

A grid wall is between private living area and the kitchen, which both separates these areas and creates a close relationship between them.

Finally, for better introducing the plan, the sketches of interior and exterior spaces has been depicted in later pages.

As can be seen from the pictures, the primary spirit of the collage have been completely preserved in the space.

The reception room space has a beautiful and panoramic view toward southern jungles. Living room space is some stairs below the reception room. Curved walls separate these spaces to some extent, but they can be easily accessed.

For designing the site according to main lines of the plan and topographic lines of the ground and site territory, a graphical and abstract plan has been drawn and the site was finalized by defining and incorporating the necessary function.

Small trees (nearly 2 m high) are planted in the northern part of the site near the pool that bar looking from outside the site and pool, but the outside view is not restricted because of the slope of the earth.

In the eastern direction, high conifers, creating a special perspective in the space, create beautiful walls. There are a number of weeping willows with specific distance planted between this wall and the building and there are relaxing beds between these trees.

There are two joint pergolas in the western direction of the site having bushes of red flowers that define the parking. The green color of trees with black background beside a white volume in the red sun light create a beautiful combination of colors.

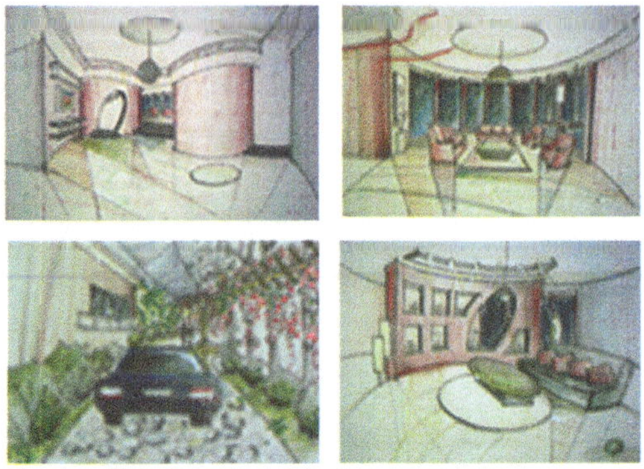

Sketches of interior space and the site

Various stages of designing a site

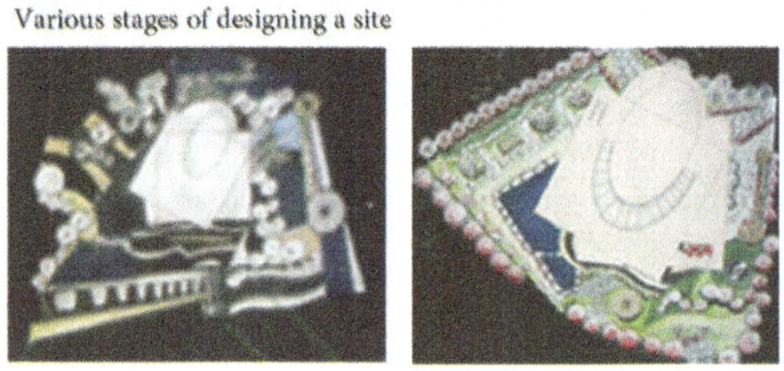

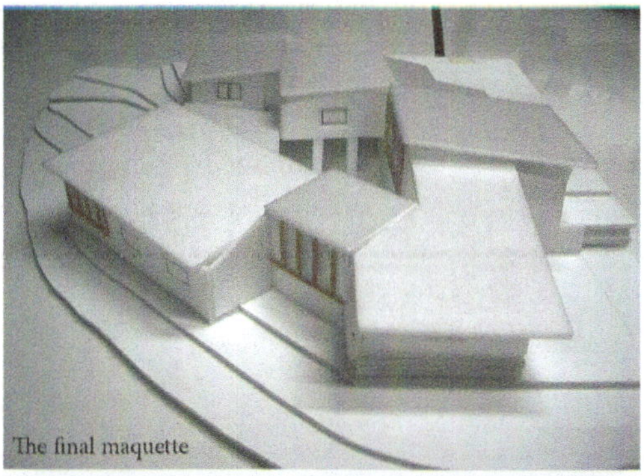
The final maquette

The first collage

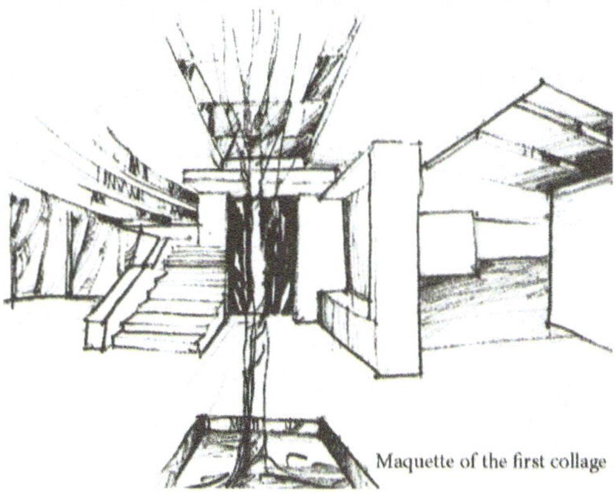
Maquette of the first collage

3.4 "Collage", a Novel Dialog

Design of the architect's house
Student Anahita Khodadadi
Masters Dr Ali Asghar Adibi
Dr Ahmad Ali Farzin

I had only the experience of one work. I accidently put volumes side by side. By "conscious playing of light volumes", I combined things and finally matched them with the informational set obtained from field and library studies and regulations and standards. To make the space pleasant, I imagined myself in the obtained volume. Some images and routes I liked to see in the framework entered the concerned volume hardly. However, I knew well that these images have hanged my volume artificially to enliven existing spaces.

"Collage" was a new dialog. Due to its freshness, my mind did not trust it and I hoped to find a better way and a more appropriate answer.

I did not want my house, the house I lived in my dreams, be merely a sculpture for the external environment. I should have lived most of my life in its interior spaces. It seemed rational to prioritize "the hope for finding a better answer" and test it like new experiences.

The road had already started.

I found my favorite images in books and magazines. I would collect those I liked respecting its shape, contexture, light, and proportion and started my early experiences. I feared. A fear of the wrong way taken. Maybe the result be meaningless. Anyway, I would lost nothing …

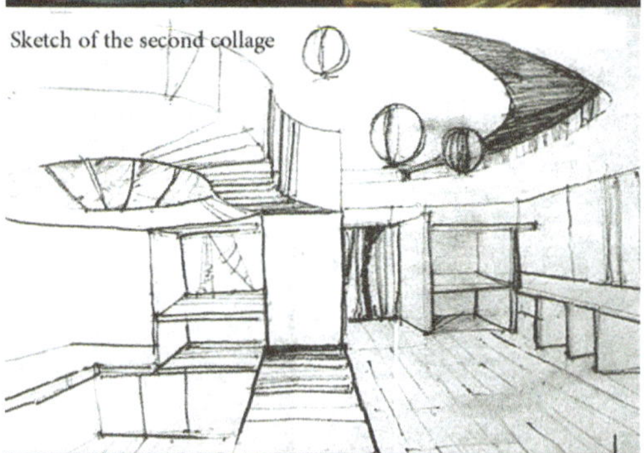

I advanced in doubt. Any piece I put beside the other and looked at it from a distance, ambiguous and vanished images enlivened before my eyes. I do not remember to what corner of my mind these images belonged. When the first collage finished and I looked at that set, I remembered my childhood house with that stout tree inside it, the branches had bulged out of the roof and its trunk had passed through the patio. I had seen it in my childhood pictures because I was too little to remember it in a corner of my mind. An open air in the warm air of the house toward summer mild breeze and hugging the external nature could be interpreted easily from it.

My fear vanished.

A set of cold and hot motions and colors and there being intertwined were the answers of the second collage. The cool breeze having dessert upon the fervent soil skin. A feeling that belong to my childhood left behind the past memories. This time I should searched the twists of my mind a little more.

Third collage

Fourth collage

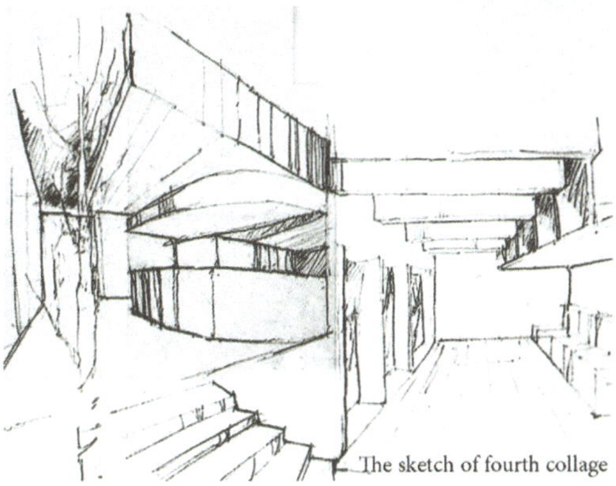
The sketch of fourth collage

The third experience was an intertwined spatial experience creating an interesting framework. A helicoid upward motion I was standing in and, in one hand, I beheld the warm and illuminated space of the below that called me to itself and, in the other, a porch faced to the periphery nature forming a corridor toward the infinite space. It seemed that the twists of the coil had opened a window in the heart of earth that could be the place for my childhood plays. "Elf I the heart of green valleys". I remembered the days in which I nurtured such a space, in the intertwined spaces of collages I shaped my dreams. This road was to arrive at the plan of a house ....

In the fourth collage, I saw the warm space of a family. A space for gathering and relying on being safe against the invasion of the cold external environment. I had to define these spaces for others.

I sketched.

My imagination were illustrated. There were no light. There were no subtlety or coarseness. I created the maquette of one of them. I enlivened my imagination. The analysis of collages meaning, not the consisting elements of them, as well as constructing a volume of them, creates spaces constituting every corner of my house. The issue of moving in the space, inside view and scene to outside, the intensity and light direction of the sun, light reflections and fractures, shaped building volume.

"My house" was shaped based on the inspiration of the same upward spiral movement that creped on the ground and the same crack that took nature in.

Then it was necessary to consider the function, standards, proportions, and other study data and correspond the plan with them. It was very interesting that there were little need to change and convert the data and it seemed completely rational that my unconscious mind identified appropriate and correct proportions and automatically implemented them in the existing volume. An appropriate answer was found for the ground and volume. Preparing the maps of buildings, sections, plans, etc. were conducted in the course of project design.

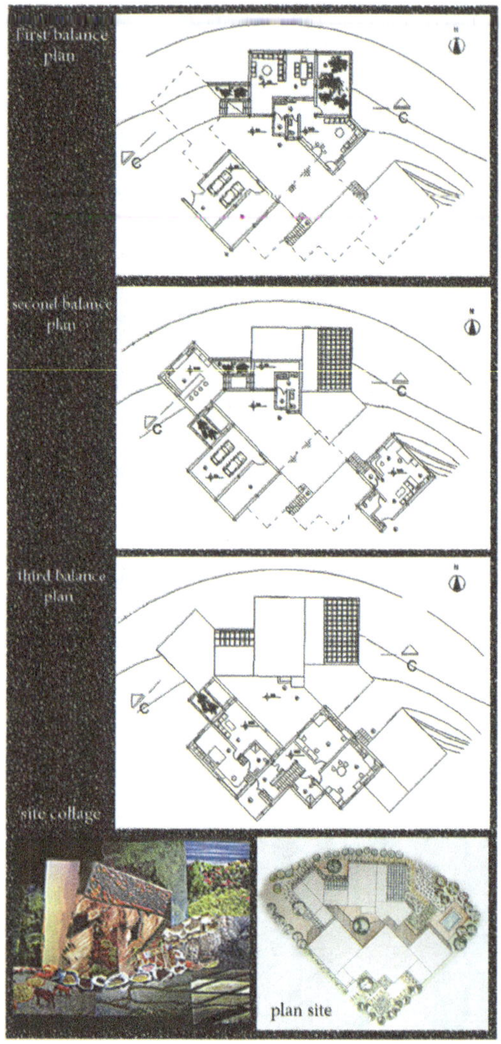

Site project design based on the plan and only by considering the issue of function, view and scope, sunlight, and the position of openings, give a correct answer, which is the only answer to this problem, and maybe the picture of "The Favorite Garden of My House" does exist in it. Though the collage of the site did not exactly emerge in the plan, its properties of contexture and material connectivity, plant coverage, light direction and colorful stains in the green background were influential in designing complex site.

3.5 The Forgotten Details

Design of the architect's house

Student	Hatef Ghahrodi
Masters	Dr Ali Asghar Adibi
	Dr Ahmad Ali Farzin

In this abstract, I introduce a unitary plan in line with concepts such nurturing the old structure and elements psychological correspondence.

In the front sample, an effort has been made to keep the unity of building elements and arrive at the interior design (interior decoration, furniture, etc.) and exterior design (total volume, facades, site design, etc.) of the building.

Architecture is nothing but the unity of space elements, because this unity influence the addressee psychologically and visually. An exalted collage helps the architect to arrive at a harmonic space.

Generally, spaces can be classified into two categories: open, exterior spaces and closed, interior places. Thus, architecture design emerges in the framework of these two types. If in designing with collage a picture of both is presented, a picture with two aspects emerges.

The following explanation are something more than word play. The obtained pictures are the basis of concept formation and in some ways are considered as the starting point of architecture design. However, remember it could be of great help when starting the design and the course of it. In other words, the designer starts with the collage and ends with collage too.

With repeated sketches from different aspects and prevalent colors in the collages, we arrive at a color key resulted for designer's mind. We identify our ideal colors in collages and detect designing route based on these ideals and recognizing ourselves. The results of the first five collages, which were intended to be experimental and could be understood from the primary designing trend, are used in shaping the spaces that could form the primary material of the volume.

Space 1-1

Space 1-2

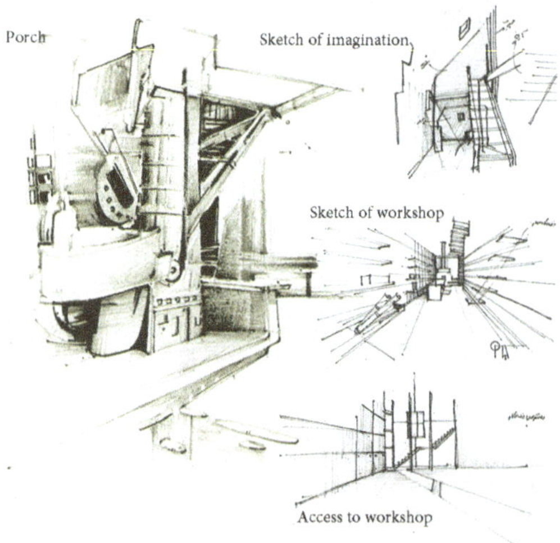

Shape and form

These meanings corresponded the concept origin, which is the same as reconstructing or nurturing culture and tradition after deconstructing indigenous cliché structures so that the humans see themselves in the previous spatial and temporal dimension when faced with a new issue. In the unitary design, geometry in the structure, abstraction, logic, and order in the shape and form of the offered products, the forms perform in a logical correspondence.

Ergonomic and function

An effort has been made to arrive at some kind of psychological and corporeal correspondence. All building elements are integrated in the design and create an overall unity. It is the very beauty and unitary.

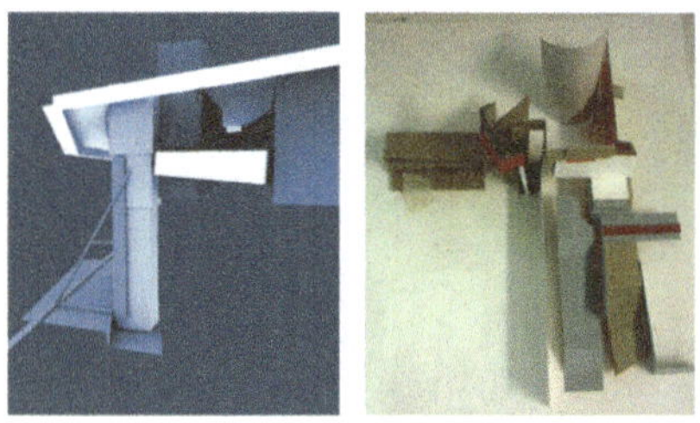

Space 2

Malovic believes that "the art was a subsidiary and excessive element from long before, permeating into the life and thought of humans when integrated with other elements. Now is the time of independence. We should learn how to read the forms". The majority of unitary building and designs were focused on balanced, glorious, souvenir structures and were designed as relies in respect of using geometry and technology.

It should be noted that collage is closely related to these meanings, because both are involved in a unitary and dynamic game. Like me, your undoubtedly believe that one should search a background in when designing. The history will not be repeated but at least it could be an example.

The first collage

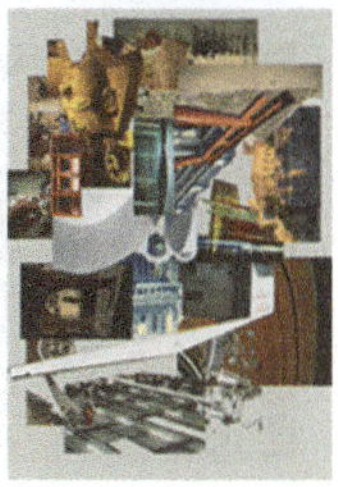

The third collage

The second collage

The fourth collage

The first collage

There are columns in this collage. The overall combination is completely systematic and implies a reception or living space.

The second collage

The colder and less balanced spirit collage compared to the first collage, implies the illuminations of an open space. In the left of the picture, there is an unstable column. Deconstruction order is evident in the left half of the picture. The organic stairs in, which is in harmony with the scene and the red band in front of the entrance, can be observed.

The third collage

The feeling I have about this seemingly beautiful and enchanting picture cannot be expressed. In the first plan, cold colors in the background are seen with an explosion of yellow color spectrums, which my conception of a skyward mirror, while there appeared a 10 m high wall from within the water that acted as the building sky. First, in face of the irritant winter wind and, second, in face of the western light covered with sun collectors and having a mirror-like manifestation in face of the southwest wind. (Though one of the terrace functions that has been design accordingly, is to behold the sunset when having dinner.) One of good elements and movements of this tableau is the fractured longitudinal movement of the lower one third.

The fourth collage

This required more thinking. The glaring light of the collage 3, the sense of a yard with strange decorations and a deep groove on the ground intersected in some points, the red and yellow lights of the right-hand window, and finally the illuminated holes of the first collage, recalls landscape illumination. Despite this expression being a little ideal and having distance from the reality, I was confident that the difference of my goal with these details was not large and tried to consider this difference only in reductions so as to not hurt the spirit of collages. In conclusion, the same way as the design result is hidden in the heart of its process, I think it is not necessary to make conclusions of this text.

The Maquette of concept 1 while designing the form, the identity of collages (the first five collages) I enjoyed its hue color (especially in the plan).

I discovered an important thing that more than the pure form of the building indicating the unity of the elements, human spaces should possess a kind of psychological correspondence.

The combined concept resulted from the last four collages, reaches the third concept after analyzing it with the diagrams of the site and topology. I the final nose (living room), various directions could be considered based on the general structure and in accordance with the regulations of the site. The confrontation of unity beauty and rational diagram leads to a solution like the concept 4. The confrontation of unity beauty and rational diagrams leads to a concept like concept 4. This concept is the result of passing from function and believing in tis example that sometimes function follows the form. Movement is one of the main organs of unity. In this structure, building is as a factory shapes the form of external dynamism through its internal actions and behaviors including life flow and daily affairs. However, its contradiction with the environment is debatable. In the next stages, I tried to arrive at a form corresponding with the spirit of the building that is the warm environment of the family in the first place.

The final Maquette progressing with the internal design, largely reached the ideal colors, light movements, and volume combinations in details and in macro level.

Composition of spaces
Concept of second collage

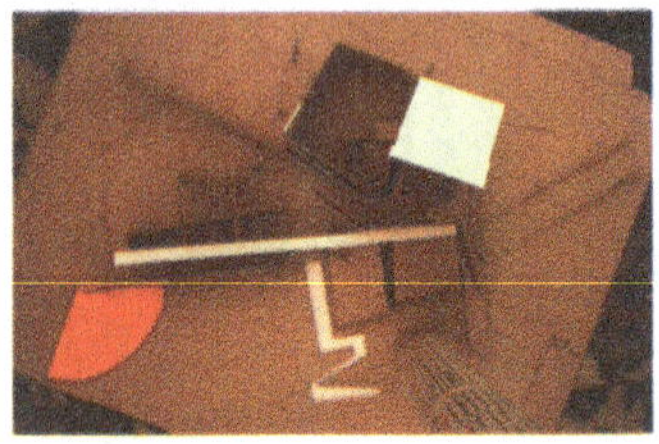

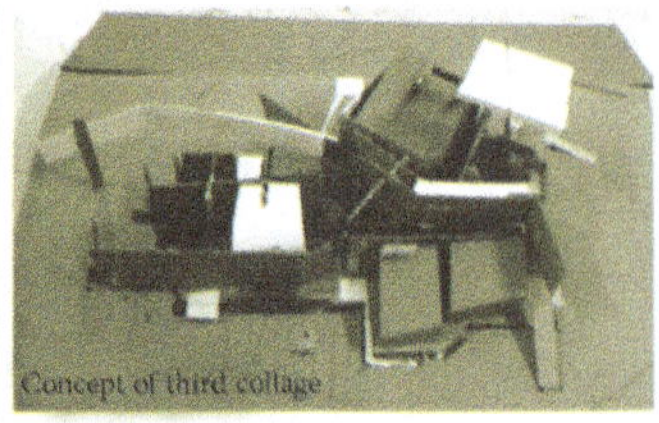

Concept of third collage

Concept of first collage

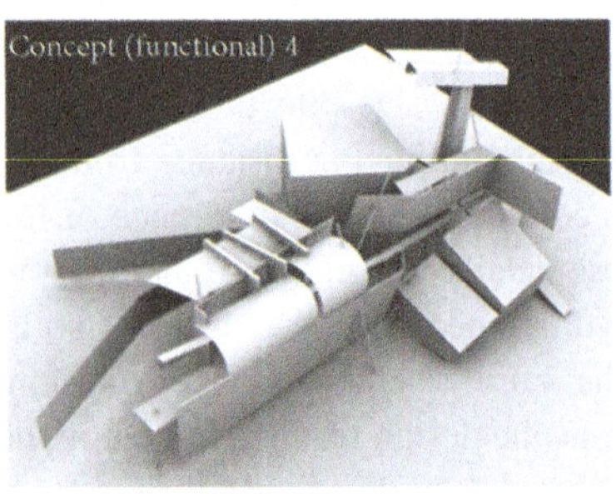

Concept (functional) 4

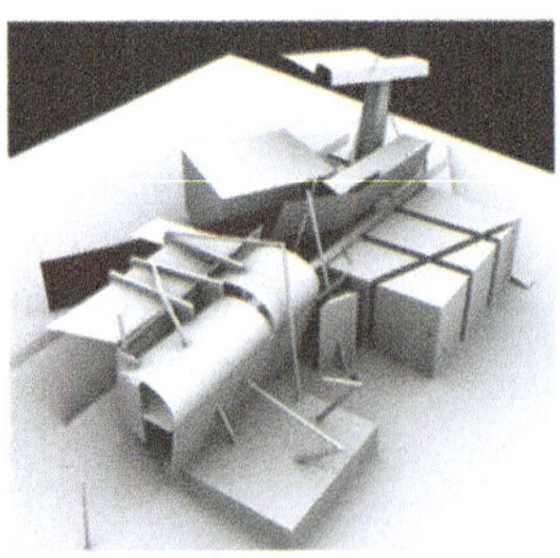

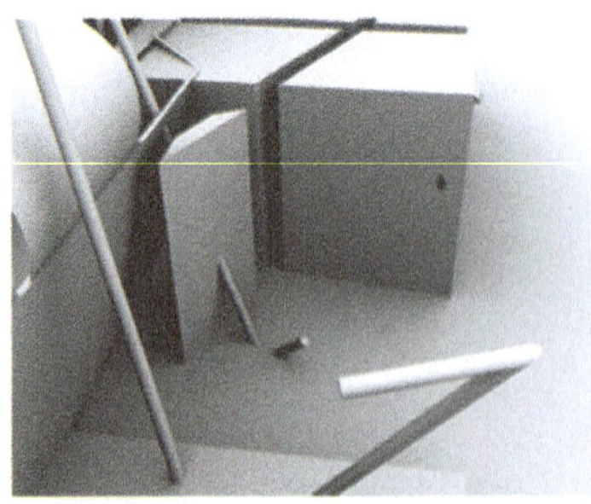

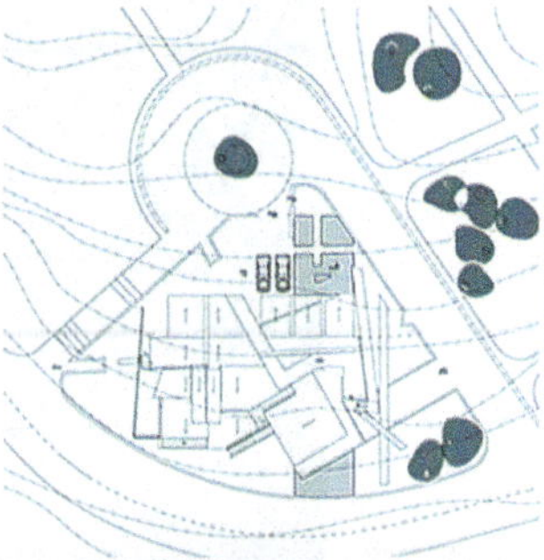

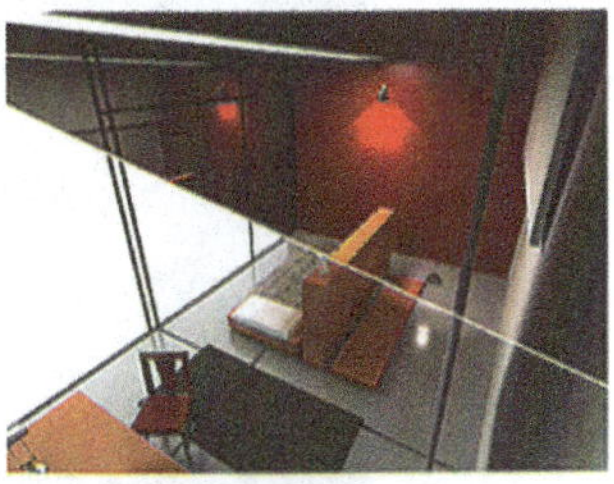

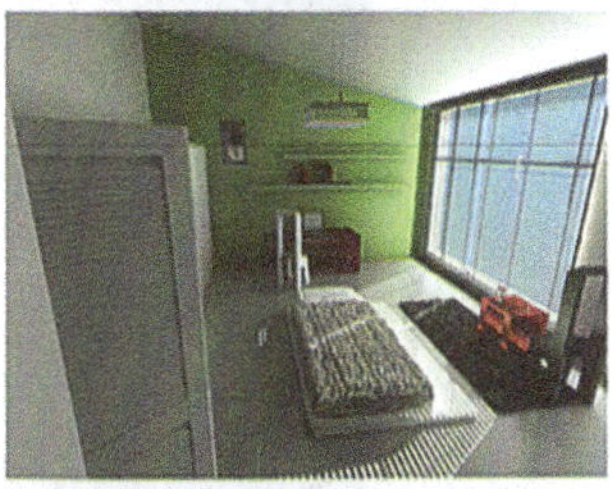

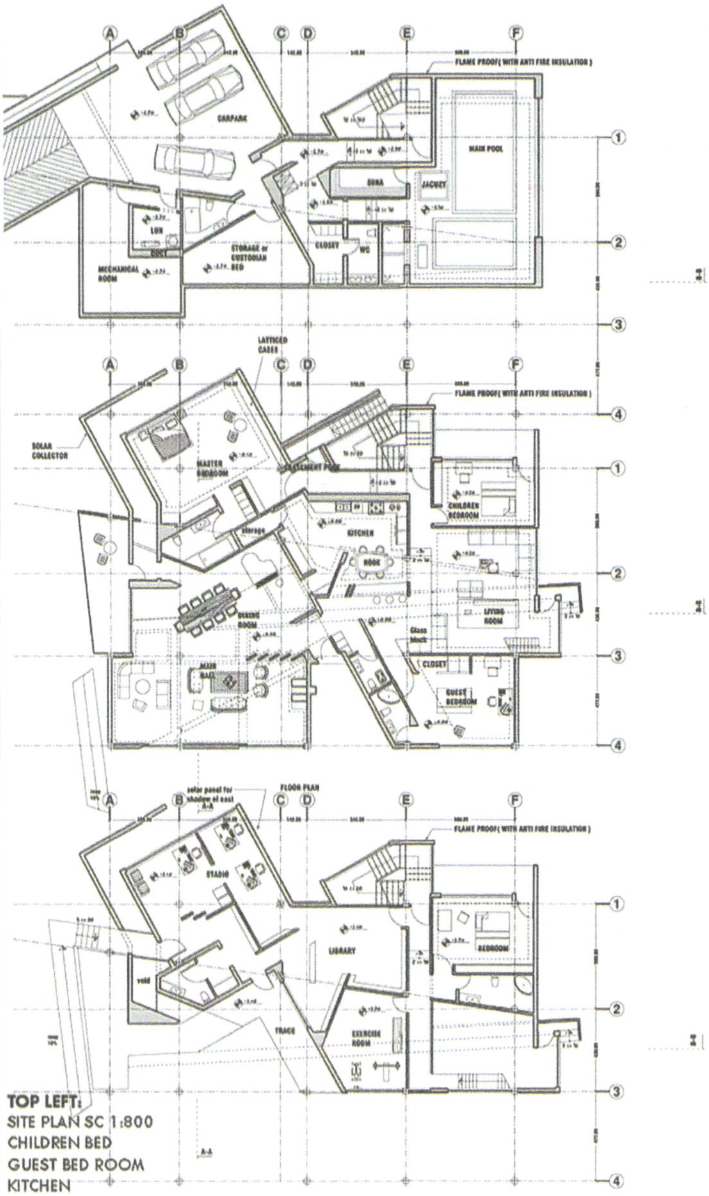

TOP LEFT:
SITE PLAN SC 1:800
CHILDREN BED
GUEST BED ROOM
KITCHEN

CONCEPT
GREEN VILLA

TOP RIGHT:
-1 BASEMENT FLOOR PLAN SC 1:400
GROUND FLOOR PLAN SC 1:400
FIRST FLOOR PLAN SC 1:400

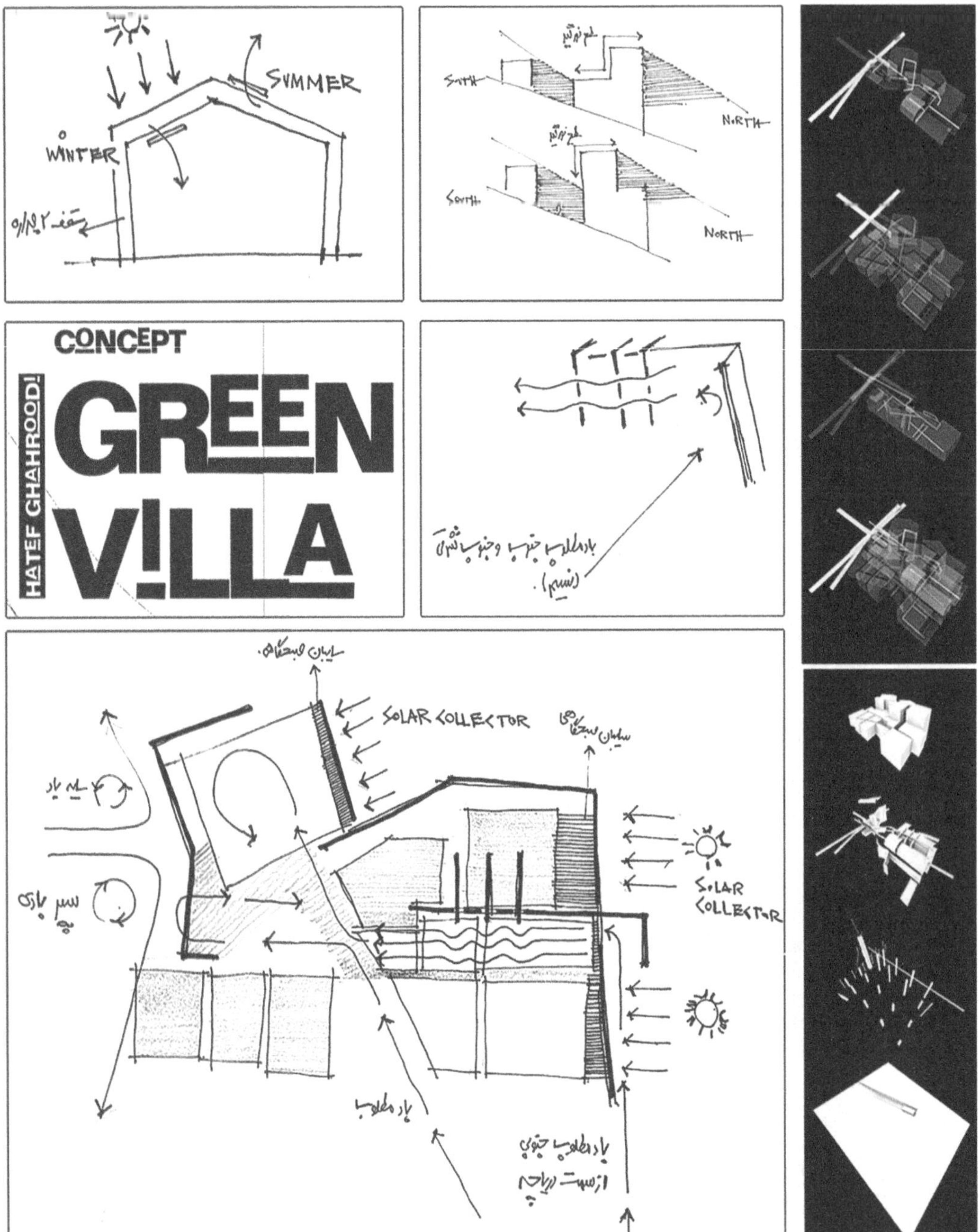
SUMMER
WINTER
SOUTH
NORTH
SOUTH
NORTH
CONCEPT
HATEF GHAHROOD!
GREEN
VILLA
SOLAR COLLECTOR
SOLAR
COLLECTOR

3.6 Involvement with Tranquility

Design of the architect's house
Student Mahan Javid
Masters Dr Ali Asghar Adibi
Dr Ahmad Ali Farzin

Passing the line, surface, reaching the infinite space, searching for an ideal place in the corridors of disturbed minds in the wastes of disheveled thoughts in the vicissitudes of the unconscious mind, how such a space has formed? Where is the origin of these pictures? An extensive world the more is sought the less is found, a new experience among thousands ways to create. Those experiences were sweet and strange. What a fictional daring it granted me. The though upon which one can rely throughout the life: it seemed that I had started a road in which more interesting events are ahead compared to other roads. Some spaces are created that do not exist in anywhere else. Some sites are created not be found except by relying on our unconscious mind. The range of these spaces are far more than other roads. A glance revealed that masters of this group are far more experienced and aware than others.

I started creating Maquette when hearing the words of masters. At first, I selected my favorite pictures after reviewing magazines and books. Not only architectural pictures, but also any beautiful picture.

I started arranging the pictures side by side. This had its root in unconscious mind and turned into the conscious mind. At first. It was hard to understand in the first place, but at last we arrived at it (does the unconscious mind assist us in reaching an ideal space?).

After arriving at the collage concerned, I fixed the pictures and looked at them at long distances. After some investigations, we arrived at spaces very close to our ideals. There were intact and strange spaces in every point.

The masters told us to hang them on the wall, looked at them, and would discover new spaces every day. We sketched according to the spaces, and it was an extraordinary experience. After some sketches, we fixed some spaces based on each word.

It was interesting: the towel turned into curtain, glass into fireplace, book into stairs, flower into hole, and boat into dresser. We started creating partial maquttes according to the previous sketches. It should be noted that the design became more complete and comprehensive in each stage to have a more tangible understanding of a 3-dimensional sense of the space.

After creating the maquettes, we generalized them to the whole space. In this stage, at first I depicted the concerned plan and arrived at the whole space with the help of integrated plan and partial Maquette.

Then I started creating the whole Maquette and afterward incorporated it in the intended site according to environmental and climatic conditions.

In all these stages, the plan became more complete.

Then, we started designing the site.

After consulting the masters and applying the changes, we began creating a more complete Maquette that was intended to be the final Maquette.

After reaching the total primary space, we examined environmental, climatic, and access conditions: issues such as ideal spaces direction, number of family members, the direction of disturbing winds, appropriate light absorption, better scenes, appropriate locations to enter into the site, etc.

Then, we started locating in the plan and places and the intended cases and created the final plan.

In conclusion, we examined building details, placement of columns, building material, etc. and completed the design.

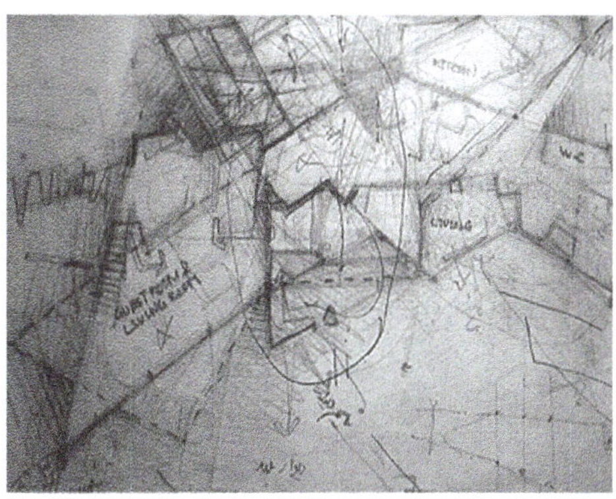

The road that had to be tested.

The road we had to test.

The road good to have been tested.

An arrangement of beautiful and various pictures, playing with puzzles, perplexity, arriving at a balanced proportion, searching for the internal utopia, consolidation on the paper, and reaching the flower and part and again a trip to the whole, solving standard problems in an imaginative design. Attaining a dynamic balance.

Solving peripheral problems in the site.

Logical and total exposition, in a superficial view only through principled dislocation and imagination on the other side.

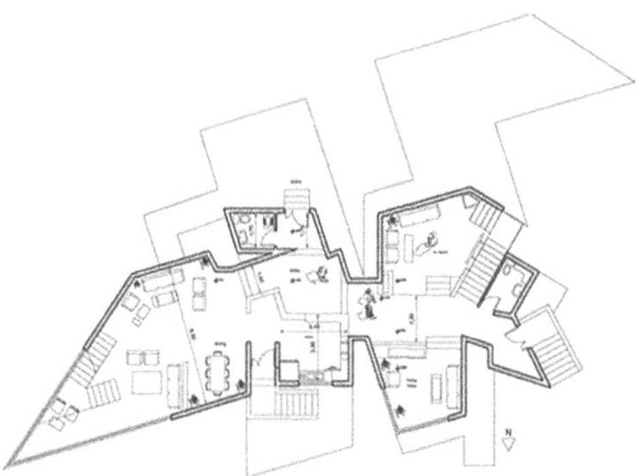

FIRST PLAN
SC: 1:100

SECOND PLAN
SC: 1:100

3.7 An Ideal Start

Designing architect's dwelling

Student	Negin Shakibi
Masters	Dr Ali Asghar Adibi
	Dr Ahmad Ali Farzin

Why collage?

During the short periods of workshop experience, which are summarized in five designs, testing different methods and working with various masters are completely logical. I always believe that we should experience various educational methods and choose the best. Thus, I selected collage as a new route for architectural design.

The first stage: seeing the collage example
Masters exhibited various collage examples and asked the students to extract spaces from them in a short period and draw their sketch. Despite the collages being common to all students, there resulted different spaces. Anyone searched the infinite world of the constructed collage according to his own mentality and view and no two sketches were similar.

The second stage: the process of creating the collage started

I put clippings of newspapers and magazines side by side to create an ideal space. With favorite colors, favorite combination of favorite parts. The result was a picture seemingly displaying the fingerprint of the creator. Collage parts temporarily had joined each other. In any moment, we could displace these parts and arrive at the favorite combination. When looking at the collage from a short distance, there were only torn pieces of pictures. However, when looking at it from a longer distance, a space within the integrated picture would emerge. Ideal, unintentional spaces were created only through our ideal combination.

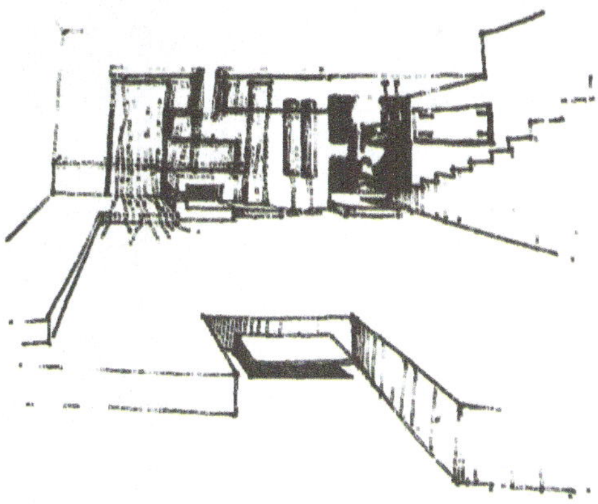

The third stage
I should make visible to the entire thing I saw. Thus, I began depicting the sketch of collages and spaces the way it should be done. The first collage included a number of spaces, side by side with a little surface difference. The lights and colors of central part of the collage implied that this space is the living room space separated from the backspace by black walls. The yellow and fire like colors inside this black wall could evidently be a warm and ignited fireplace between two spaces, heating both spaces. There is a stairs in the right-hand corner of this collage going upstairs. The black part beside this stairs can be thought of as a wall in the end of a short stairs upon which a poster has been installed. There is a part in the center of the collage through which mild green shadows along with a little light enter the space. Because there is low natural light in this space, probably the final volume exists in the west direction. The right-most corner of the collage shows parts of back-space of the black wall. The intimate and warm colors of this section can be considered a dining room and a kitchen.

The fourth stage
The Maquette was created from interior space with complete details and through using collage colors and materials of the picture. Various materials were used so that the Maquette approaches the main spirit of the collage as far as possible. The opaque plexi for generating indirect light and various kinds of cardboards for determining the genus in flooring and separating walls. The created Maquette helped humanization of dimensions and sizes and spaces seemed more real. Everything was so exact and natural that it was possible to sit in the reception room for a moment and talk with the guests.

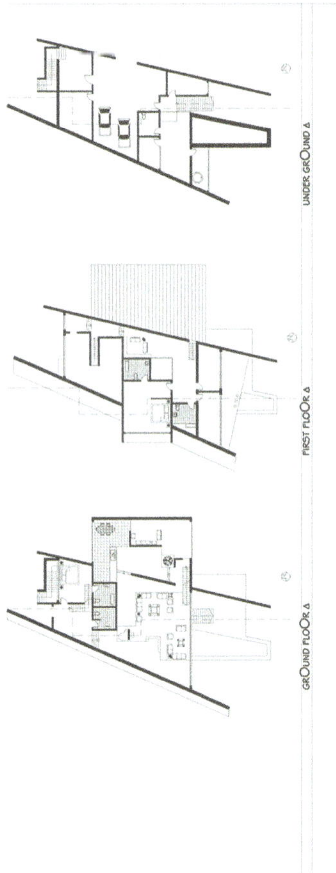

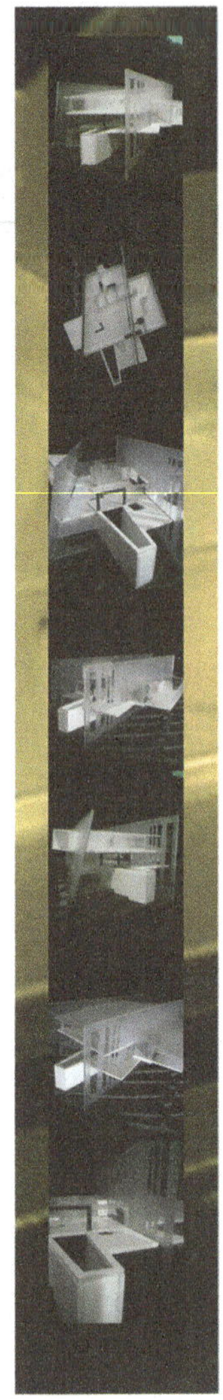

The fifth stage: designing

The starting point is one of the most important stages of designing. May be we have always asked ourselves, "should I start with the plan or the main volume or …?" In fact, collage defines a starting point for us. Using the spatial geometry obtained from the collage, I started designing. I assumed the primary space as the starting point and extended the design to other directions. The final design was obtained with a little change in which the primary core was evident in it. The locating of some functions was performed with collage method, including reception salon, dining room, entry stairs to upstairs, and the kitchen. Then, the design of a two-store house with an underground was obtained. The primary design and final Maquette was created. The Maquette contains interior details in which the primary core is evident. Thus, the collage created from pieces of my mind turned into the ideal house of the architect!

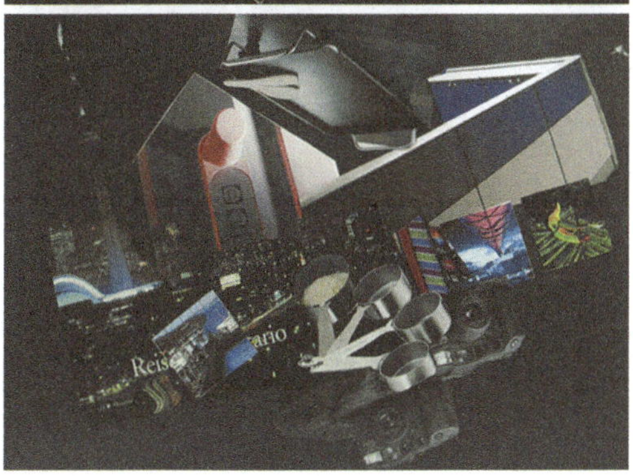

3.8 Intellectual Motion from Inside to Outside

Designing architect's design

Student	Nina Goud Asiaee
Masters	Dr Ali Asghar Adibi
	Dr Ahmad Ali Farzin

The internal substance and material have been the focus in designing this complex. Then, the external form of the building was considered. The intellectual motion has been from inside to outside.

Firstly, with collage, which is an interesting and attractive method, a part of the complex was seen.

Working with this collage was such that attractive pieces of magazines were selected and cut. It should be noted that no rules were followed when cutting the pieces.

The cut pieces due to having picture and color, contained light, gender, and material and constituted a 3 dimensional space when looked at from a distance. This space may be the favorite space of the individual creating the collage.

Six collages were created one after the other. It is worth noting that in the first two collages, the spaces were not very clear but in the next four collages, due to experience increase in selecting pictures, the relations of juxtaposed pictures increased and thus there appeared a space that I loved more and a better relationship was set between the collage and me.

After reflecting upon the observed sections, I tried to generalize and extend them.

The main issue put forward in this method is to arrive at whole from part.

By part, it is meant a small section of the design seen in the collage and an effort has been made to preserve the main part and thus add the other sections. It should pointed out that by using other collages and creating a logical relationship between them. It was tried to juxtapose other sections through the defined relationships.

The collage helped to the progress in each point and guided it and in some ways has been the leader of this inside outside motion.

In all the designs in which this method has been employed, there is a close relationship between the design and the collage, which implies that this method is very rich, analyzable, and worth reflecting upon.

Even illumination and material have been completed using this method.

The early sketches depicted according to the collages

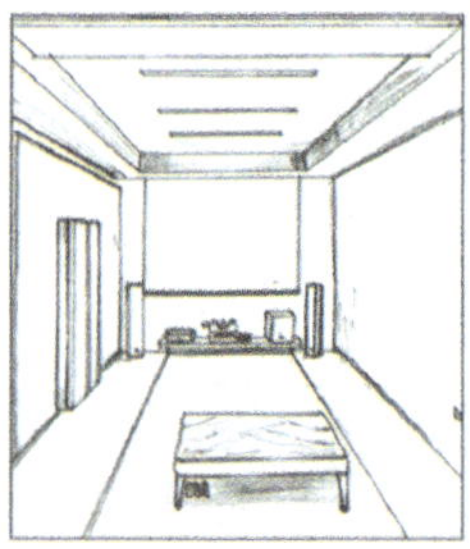

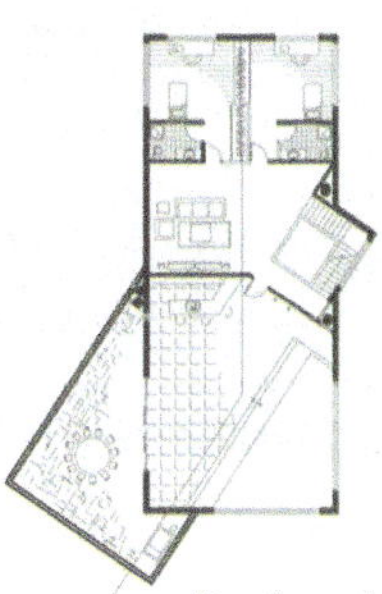

First floor plan

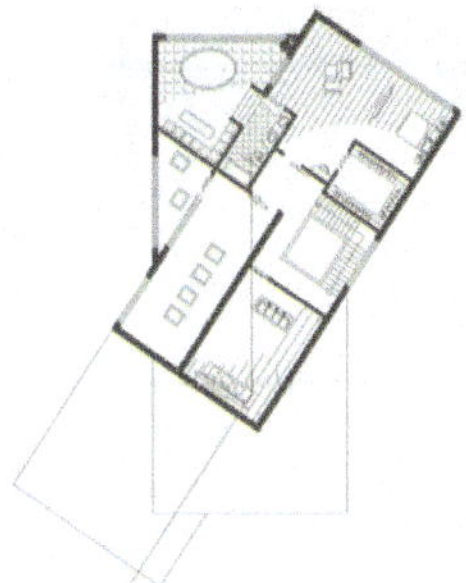

Second floor plan

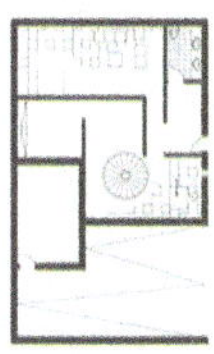

Underground plan

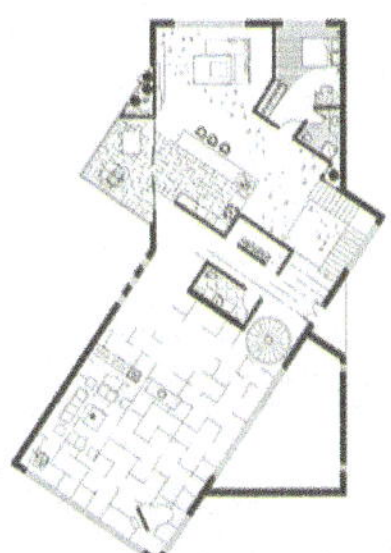

Ground plan

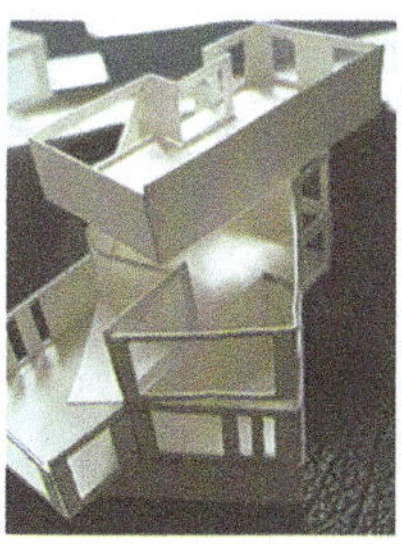

In this interpretation of this motion it should be pointed out that, each stage progresses gradually with the collage. Albeit, after seeing the space in the collage they will be sketched. Then, a Maquette seen in the space was created, leading the creator into the space. Accordingly, the plan related to this space was designed.

The important point is that the concerned spaces were developed and sketches and maquettes were made use of to advance the plan.

The best efforts have been made to define spaces side by side with the help of the collage. The sketches and maquettes created situate the addressee inside the space.

Internal sketches and maquettes and inside outside motion depended upon the plan program, because it was the issue of residential house design in which the most important issue of the house interior was the relationship of internal spaces and illumination that we considered solving the most important issue using this method. In other words, we deemed the issue as solved and entered inside the house and designed the house based on comfort, ease, and the access of individuals. Then, we started illumination and when interior spaces relationship was solved, we considered the exterior spaces.

The material tangible in collages, were employed in the façade. Tall windows were used to provide internal light.

Due to the warm colors of the collage, tall windows were used in all defined spaces as far as possible.

After the façade, exterior spaces were designed. The ways to access the building and the relationship of spaces were observable when deliberating upon the collage. In fact, the collage was a sparkle that finally facilitated the access between spaces, negative spaces the so call full empty spaces.

In conclusion, it should be noted that in designing a complex the focus of which is the user, all conditions should be considered and in fact, inside outside motion is the best solution to arrive at the best result.

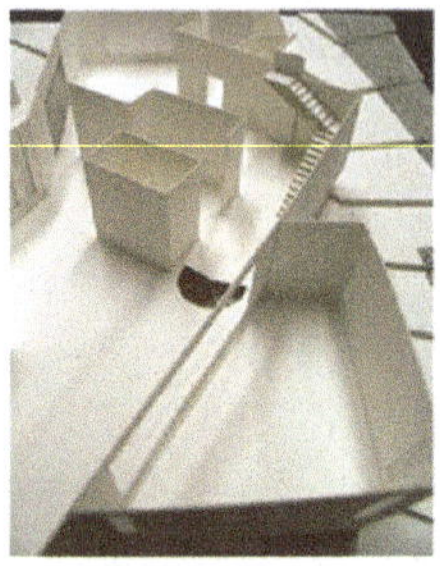

3.9 The Sparkle of the Mind of "Collage"

Designing architect's dwelling
Student Ali Sorush
Masters Dr Ali Asghar Adibi
Dr Ahmad Ali Farzin

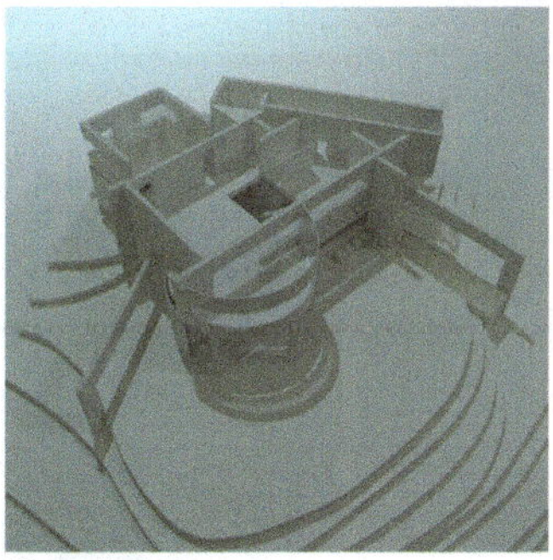

Above: final Maquette
Below: project collage

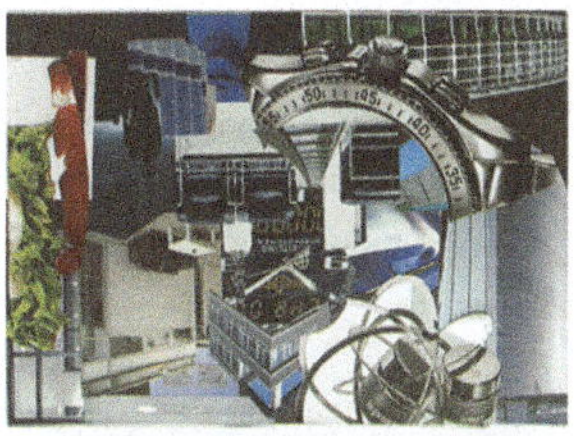

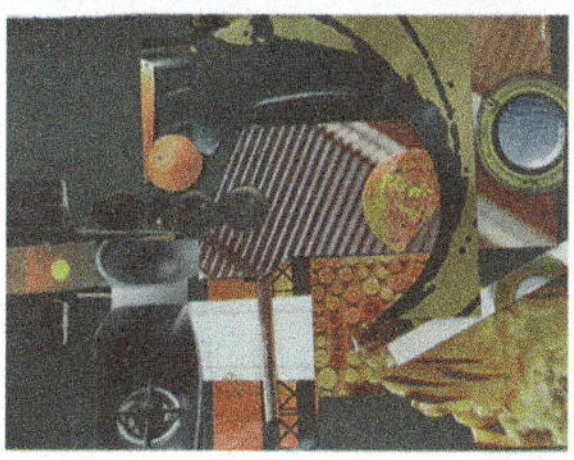

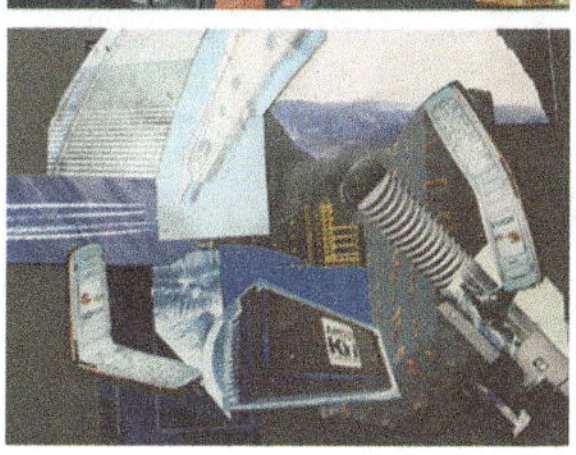

When I entered architecture atelier 2, I thought that I should pass the new design like other designs, i.e. based on a general process call architectural design. For doing this, at first extensive studies should be conducted about the site and climate as well as cultural and architectural studies. Then, all information, views, and tastes should be imagined as concepts and be drawn on the paper. Afterward, the concepts should be edited during the time period intended for the design and implement it.

The main problem in designing using this method was actually its momentariness, that is, the method being based on mental sparkle that could not be all the taste raw and comprehensive data of the designer architect.

Collage is arriving at a general concept, space, concept, and idea being an unconscious range of momentary process. In collage, favorite picture clippings are side by side through a spiritual order, and the landscape of this work is to create a space with regard to micro-tastes and forgotten details of designer's mind.

After it is created, then it should be looked at. A look with architectural and creative eyes seeing the context side by side, color and form as the raw material of designing, which the concept. Then, it creates this space in the mind and its Maquette was created for transferring mental meanings and stabilizing them.

The Maquette that expresses a feeling and is different from the regular modelling. In this Maquette, the architectural mind simulates the space and contexture.

A Maquette, a space!

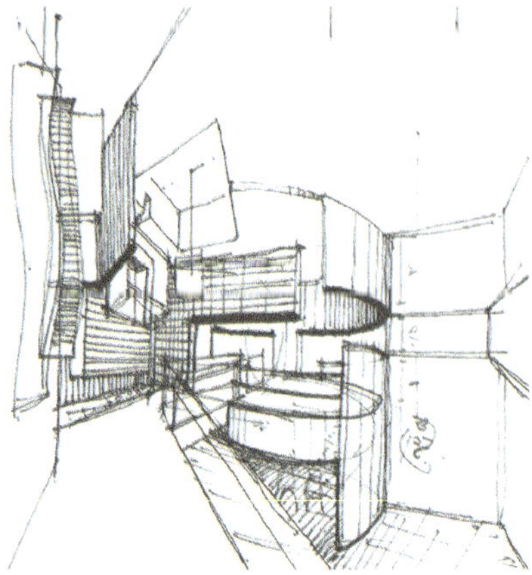

Above: a sketch of the collage
Below: pictures of collage maquettes

Above: collage Maquette
Below: collage plan etudes

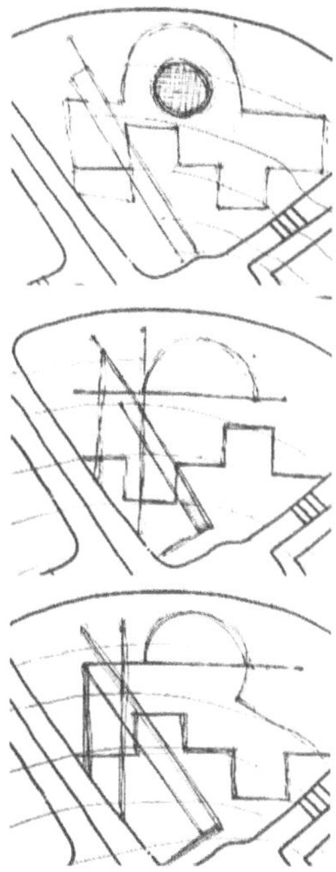

The next stage is to convert his landscape into space and volume. This method that accompanies creativity during the process of arriving at the design, considers flexibility in work stages as its inherent principles.

Selecting the size, form, and the cut graphic, and placing it in the intended position, and then after creating the collage, the manner and view angle as well as sketching, are among the issues available for the designer to create a work encompassing all thoughts and interests.

The development of the same mental sparkle or the same collage in this section that is the concept and space, is completely different with the general method. General principles are the same. Complete studies have been conducted about the design in the area of site, climate, culture, and civilization as well as extracting a concept out of the collage.

One of the very useful methods is to edit and reform the concept and the maps of Maquette construction in each stage as well constantly observing the space created.

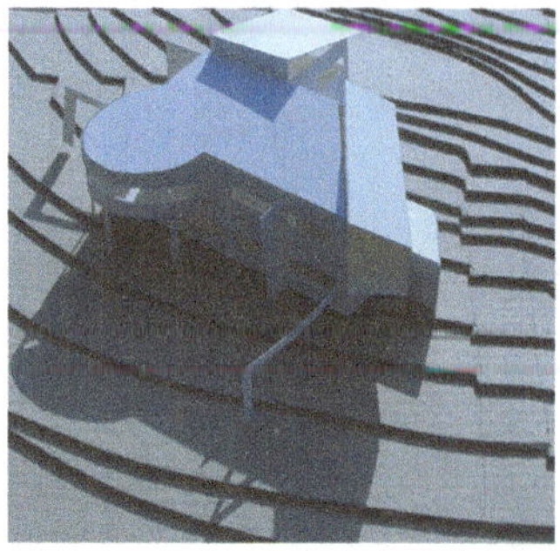

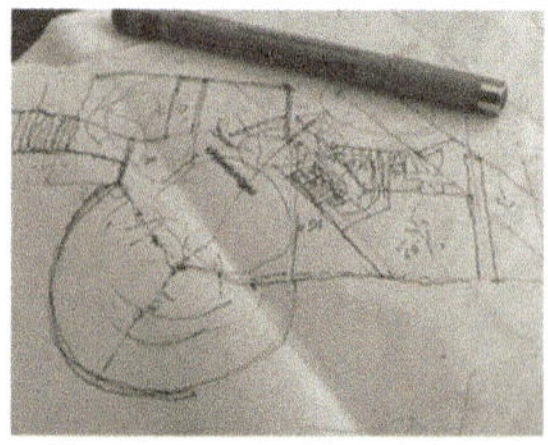

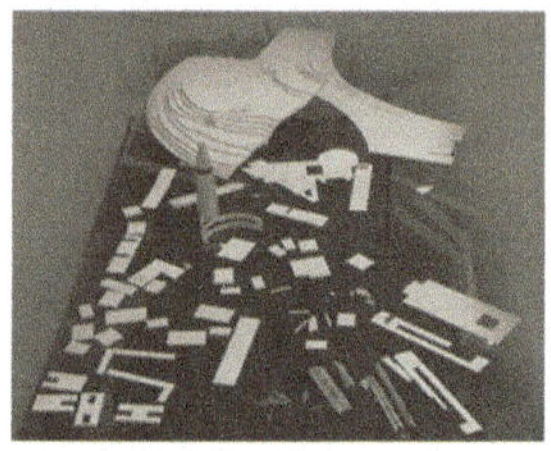

(Top-down)

Collage Maquette and project progress stages

3.10 From Inside

Designing architect's dwelling

Student	Farhad Binazadeh
Masters	Dr Ali Asghar Adibi
	Dr Ahmad Ali Farzin

Above: final Maquette

Below: first collage

At first, collage was a new experience to arrive at the design. The interesting point in this method was to reach something that forms within the humans. The method, which was possible in this design that was a residential house, was the method that could not emerge in other methods to attain the form and design of this issue. Because in working with plan geometry only 2 dimensional spaces were created, which were very vapid; but in this method instead of using 2 dimensional spaces, we had pictures of our favorite house, which the space was obtained in addition to the plan. As mentioned above, the interesting result obtained using this method, is that all the pictures created using collage, though simple in the first glance and for the common people, due to selecting colors of the magazines by the individual, it was possible that the color of spaces or even their being introspective and extrovert be pointed out.

For example, using natural colors such as blue and green is indicative of individual's inclination toward extroversion and even in the next section, i.e. illumination, it is evident; so that the color of these lights showed that whether the light is natural or artificial and internal. In case of creating spaces was to displacing the surfaces, so that all the cases mentioned were visible in the collage from a distance of three meters, so that even in season change little change occurred in these spaces. The interesting point in the collages is that even in the longest distance the space can be observed in the collage.

Due to unfamiliarity with the new method, at first, the constructed collages did not help creating and designing the spaces, I had in my mind. However, the same collages because were the result of selected elements, could be of help in choosing the material and light, so that from these early collages, the extroversion sense and the inclination to have a high view toward outside was evident. Every piece were place in a location after being cut.

Above: second collage
Below: third collage

Above: fourth collage
Below: fifth collage

All in all, when looked at from a distance or at least at from 3 m distance, due to the juxtaposition of paper clippings it seemed the lighted shone into it from a certain point. In the next stage, I created a collage and in the end, I saw that unconsciously the favorite space has been created and is pleasant for the collage creator. One should be aware that magazine clippings that create a pleasant space when put side by side that should be thought of as one of the spaced intended by the collage creator, so that other issues such as light and material were determined to some extent, have obtained this collage.

The next section was to sketch the collages, during which the spaces had to be drawn on the paper and draw plans of the sketches. In fact, part of the favorite plan of the building has been obtained. In this stage, an internal Maquette from the same section was created, so that one should be careful even when choosing the material of this Maquette and here the material are highlighted as one of the important issues.

Other important parts of the building can be created with this collage method. I created the main part using the collage, the first part of it being the reception salon; so that there are external light and wide windows with a southern view to the nature. Here is the point where the building starts. The relations of spaces were created like previous methods, so that buildings were placed side by side using various relations of these spaces using the collage.

When reflected upon, this method can be considered an internal method to arrive at the intended plan. Selecting this method was at first for testing a new method to obtain the architectural design, but it can be deemed an effective method in reaching the design using spatial view and without 2 dimensional paper depictions. Because from the very beginning there are 3 dimensional perspective and internal Maquette, the design which could be worked upon using the collage are those designs in which external volume of the internal space is less important, the residential dwelling is the best instance. In conclusion, it should be pointed out that maybe using collage method to get the dwelling design be more appropriate than other methods, so that the method of solving relations in this design using bubble diagrams, etc. is as the previous methods. Using the collage is in fact using ordinary methods for solving relations along with a spatial view of the space.

Right: final Maquette

3.11 Searching for the Favorite Space

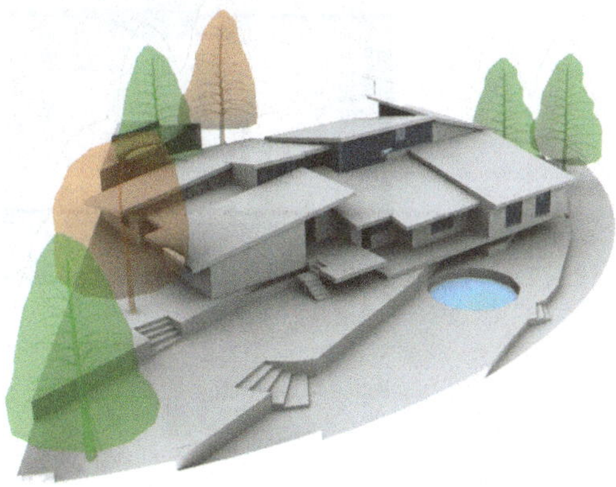

Above: project simulation
Below: the collages

Designing architect's dwelling
Student Laleh Reshad
Masters Dr Ali Asghar Adibi
Dr Ahmad Ali Farzin

In my opinion, collage is to visualize the spaces maybe hard to draw. Collage is to look at and search favorite spaces carefully amongst the pictures and photos we love.

At first I considered the colors I loved, such as chocolate, brown, and yellow, each indicating certain material, like wood, brick, and colored glass.

In fact, it was to find dark spaces of mind among favorite spaces and arranging them.

Sometimes beautiful and similar spaces were created from various spaces; i.e. between those spaces, I wanted to have more options.

I looked at the collage with a multi sketch view, so that when I combine them, there emerges a more interesting form.

I combined the spaces. The combination of spaces and sketches created a new space from a longer distance.

I planned to create a collage for each design stage.

At first, I thought no results would be obtained because it was the first time I drew using this method. However, after observing different pictures and arranging them side by side as a collage, new sparkles hit my mind.

At first, I created collage for interior spaces; from the furniture to interior spaces and various material to be used in interior architecture. Then, I started creating various green spaces, which there appeared interesting landscapes. Then, I constructed façade collage, which helped me in selecting the material and roof slope greatly.

The first collage
This collage is composed of various forms including curved and cubic spaces that imply a general cubic form. Of the spaces seen in this collage are corridors ending in a warm salon and an intimate family space. A house summarized in a single story. Before using collage method, I liked half-story states of the house. When I created the collage, the juxtaposition of spaces and their combination implied half story states. Moreover, I could use various forms of spaces and various light states of spaces in the architecture of the house. There were wide corridors facing entrances and separating spaces in the collage.

The second collage
The building façade is used to construct the façade, implying a stair building with various slopes. Materials such stone, brick, concrete and clay are could be seen in the collage. The stone and bricks used in the façade and roof clay have been obtained from the collage.

Above: first collage
Middle: project site
Below: second collage

Up: Sketch of collage
Below: Model of collage

The third collage

The third collage was the start of site design. Natural forms, green trees, and pavement were of the results of the collage. The stair form was used in the second collage and it was in turn merged with site topography.

Design process

At first, I created the collage when seeing various pictures and combining them. It was difficult to select the color and space arrangement.

When I completed the collage, I carefully searched all interior spaces to find the favorite space made from merging the pictures. I sketched based on collages. As the master recommended, I dedicated about 3 min for each collage to draw exactly the same space. Sketching the collage was like tripping in the heart of collage spaces, so that I felt I have stepped into various spaces of a house. In these circumstances, I only searched for a common aspect of these spaces.

In the next stage, I began creating the collage Maquette because I'd seen the collage and know its different spaces. Only through constructing it, there emerged the routes connecting the spaces, the result of which was half stories next to each other.

After creating different maquettes, I obtained most of the spaces I wished. Then, I create façade collage to advance exterior spaces along with interior ones. The stair shape of the façade collage followed half story form of my house. These similarities in collages were interesting to me.

After implementing functions in the plan completely and incorporating various spatial links in it, the primary plan took form, which was only a single story. Then, I considered making the plan beautiful and I changed slope forms and their angles. Thus, I designed the plan along with the façade.

Among the issues added later, were determining the height of half stories, creating greenhouse spaces and different entrances for different parts. When spaces were created completely, I finished the slope of the roofs, a separate slope for each part of the building.

Then I created the collage and considered various views and scopes of the site. I tried to use site topography for designing it. Short tress in front of the building did not bar the view toward the city.

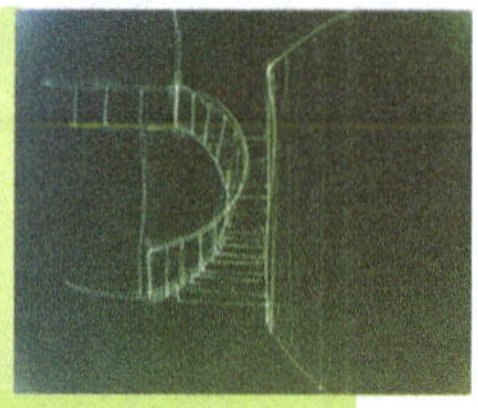

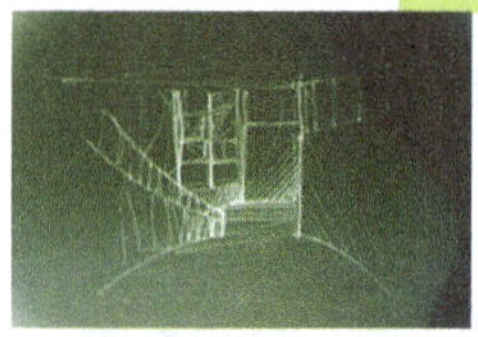

Above: a sketch of the collage
Below: collage Maquette

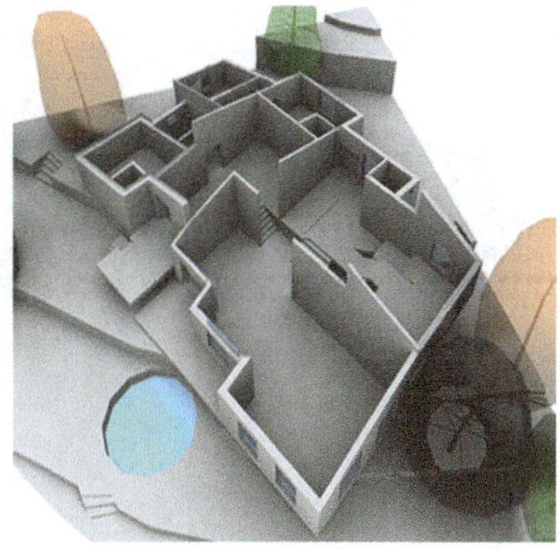

Plan

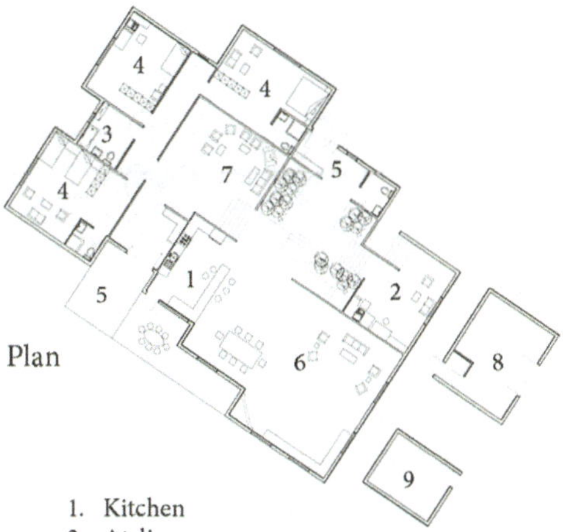

1. Kitchen
2. Atelier
3. W. C
4. Sleeping room
5. Entrance
6. Living room
7. Private living room
8. Parking
9. Installations

Interior space

Facades

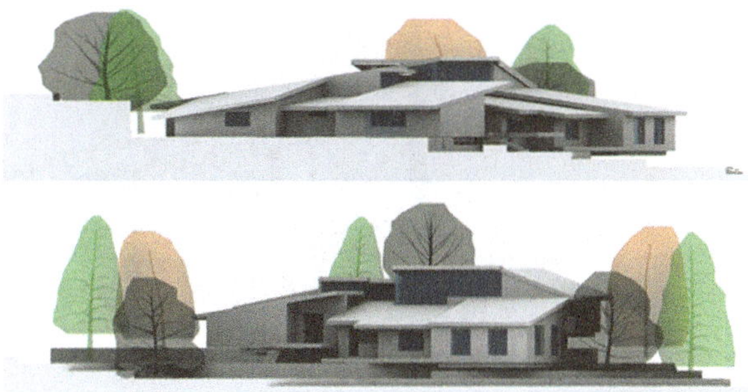

Site plan

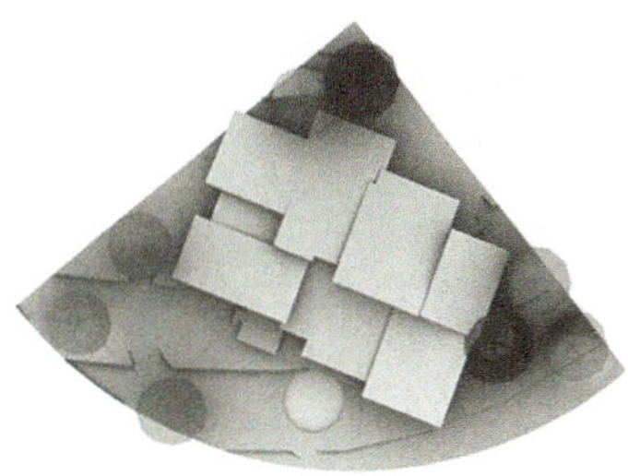

Slope of roofs

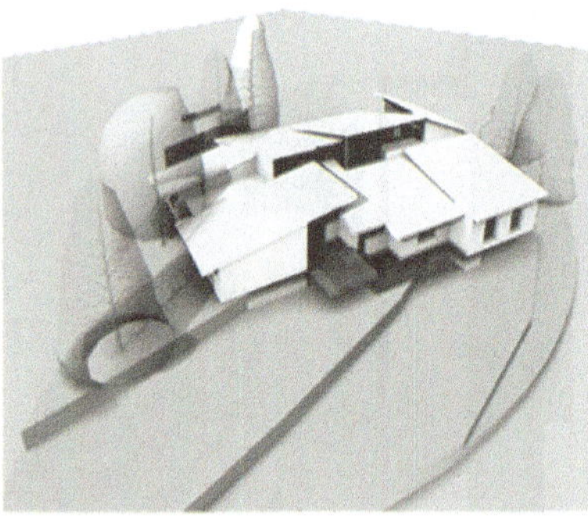

Parking

The framed picture

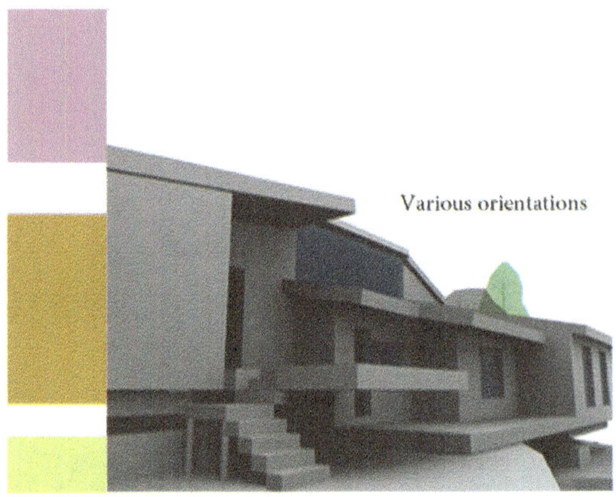

3.12 Recovering Spaces in the Hidden Corners of the Mind

Designing architect's dwelling

Student	Niloofar Zarnegar
Masters	Dr Ali Asghar Adibi
	Dr Ahmad Ali Farzin

The start of collaging was the beginning of a new experience. The first time I started exercising collaging, I felt it is an unprecedented experience. Cutting clippings of newspapers and magazines, pieces in which certain characteristics such as light, combination, spatial sense, etc. attracts the attention, and then arranging these clippings next to each other upon a black cardboard ….

The exercise was at first a bit ambiguous but interesting. When turning the pages of some magazines, it seemed some pictures called me to themselves. Maybe I could not say that what characteristic of a certain picture attracted my attention, but it seemed the pictures were selected with me not being aware of it.

After cutting many pieces of these magazines, I entered into a more critical stage to merge these pieces and arranging them on a black background. It was difficult to select the first piece. I tested a number of pieces on the cardboard and finally the first piece was placed on it. In a distance from the picture, one could see something, and when approaching it, there appeared something else. It can be likened into a kind of "tranquility" for the imagination power.

The first picture was a cautious picture of "myself"

When creating the first collage, the goal was not important to me. What is to happen to this picture? Should I use the same characteristics in the final design? Can it be possible to imagine a 3-dimensional picture among the clippings of these magazines? These and several other questions were in my mind when creating these pictures. It was the reason why I dealt with the first collage very carefully. I selected the first collage after considerable investigation; the obscure picture of the organ of a human being was placed in the center of the page. After selecting this piece, it seemed that the orientation of the work was determined. The textured surfaces of the azure sky next to the white shiny textures and a golden curve closing the cadre.

The result was incredible; a cautious picture made by me. A reflection of personal tastes and my interests placed upon an A3 paper. It seemed part of the hidden dimensions of my mind was revealed.

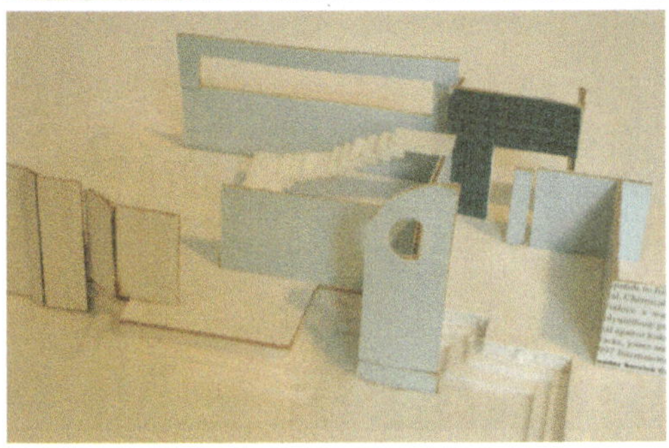

The sketch, Maquette, and 3 dimensional view
Sketching based on the pictures was not that easy at first; I constantly approached and distanced from the pictures. I tried to forget the reality of each piece. I forgot the column being a column, and the cup being a cup and forgot to use my imagination. It was a difficult exercise, but after a couple of efforts and the recommendations of the master, I managed to put on the paper my vision of the hidden space of the picture in the form of a one point perspective; an ideal space for me. For doing this, I used blue wallpaper and newspaper. Finally, the picture that seemed to be 2 dimensional and subjective, turned into a 3 dimensional and objective reality.

The fourth collage
The most effective picture in designing process

I created some other pictures after the first collage, and then other sketches, maquettes, etc. When reaching a stage in which the collage process was completely known, I decided to create other pictures without any fear or ambiguity. This was my fourth collaging experience. I selected fearlessly and combined the pieces in a full cautious manner. The result was evident. I created a completely fearless picture.

An arched, silver element was placed in the center and two spaces with different states were placed in either sides. In the final design, this was defined as guests' terrace and the spaces in either sides were the reception and living rooms.

The combination of collages and the start of dwelling design

The work site was a completely slope ground. The site maquette had already been created; I analyzed it and now was the time to put all the activities—including studies and pictures-side by side to arrive at architect dwelling in Lahijan. The project site was on Sheitan hillside and it should include dwelling and atelier of an architect. The work started by specifying reception and living rooms and their terraces northward, the direction having the best view. Then, the entrance and dividing space related to the first collage was situated in southern and southeast directions.

Then, other parts took form. The sleeping rooms in second floor, work office on work terrace in southeast and the kitchen in southwest for better light absorption and solving all relations, were other stages of the work. One other important matter was to putting this design in harmony with the slope site of Sheitan hillside, which hidden space were incorporated into the collage with regard to this slope and surface differences. Consequently, the final design was in harmony with topography of project ground.

Contrary to what was assumed to be an abstract and imaginative work, the projects were very pragmatic and functional and all relations and requirements were appropriately solved. The spaces having origins in the unconscious mind. Generally, I can state that architectural design experience was two separate experiences of a single design. Instead of previous semesters' process—i.e. considering a certain volumetric combination and then solving its relations—this time it had experience a specific trend resulting in creating more humanistic spaces. Instead of focusing on volumetric combination of the work and paying attention to bird view, this time human view and interior spaces quality were prioritized and then volumetric beauty was considered. Because work has its roots in the hidden corners of the mind, the architectural characteristics were more in harmony with human properties. For me, it was the first experience in which the process of my design was different from the process of a sculptor; a full humanist design.

3.13 Releasing from Red Lines

Designing architect's dwelling
Student Fatemeh Hashemi
Masters Dr Ali Asghar Adibi
Dr Ahmad Ali Farzin

The difference of this design with the previous one and generally with previous designs, was to select our favorite design among two methods. The first method that was the previous common method, was designing based on concept or a primary idea. The second was to design using ideas extracted form collaging and based on the spirit governing the space.

This time I decided to attain a new experience by designing using the collage. At first, I had no exact and complete understanding. I knew, I had to collect and juxtapose various pictures and finally turn them into spaces. It seemed very difficult to me. At times, I frustrated and decided to change my method but soon changed my mind.

The most important matter in this method was to obtain a spirit appropriate for the space. In my opinion, the combination of pictures along with its conscious aspect, which is the same as creating visual harmony between pictures, has also a conscious aspect that is here the harmony and cooperation between the pictures and we may say that the effect of second aspect is more than the first one when designing with collage.

In any case, I had selected this method and had to compete it. Despite the difficulties of starting the work making it difficult to spot spaces among numerous pictures, for them, but now spaces displayed themselves before my eyes. The pictures went next to each other automatically and told me how to draw. Finally, when I looked at my design, I could see my favorite tranquility in it that was also in my collages. I managed to release my mind from the impediments of red lines and obtain a valuable experience.

Living room space, close to the kitchen

The first collage

The pictures of this collage were very clear and the element of widow was used in them. The light of day coming into the space transferred tranquility and motion senses simultaneously. The spaces of these spaces are more like the spaces that require light, such as kitchen and living room as well as windows, which conduce close relationship with the nature.

The terrace near the motel

The second collage

This one is the combination of light and shadow and constitutes those spaces that need less light. Entrance hues and the terrace can be observed in these pictures.

Space maquette

Maquette of entrance at night

The third collage

This collage shows spaces in night light. Living room, corridors, entrance, etc. can be seen in the collage. Contrary to the first collage, which was enlivened by a great deal of light, this collage with its low light granted more tranquility to the night.

In the final design, the points of collages were observed as far as possible. The accesses have been considered based on the function and value of each space and with regard to design climate, ventilation and high light.

There are different sections in the designed dwelling regarding the usage of spaces including public section (guests' living room, guests' dining room), semi-private section (family's living and dining rooms, playing field for children), private section (parent's workspace, sleeping rooms). The location of these spaces is based on different factors (light, sound, access, etc.). Generally, the kitchen and family's living room and private entrance are in northern direction. Moreover, this section has a very beautiful view toward the lake.

Guests’ living and sleeping room is southward so that it is near the main entrance and has enough light. The sleeping rooms for family members and playing field of the children are in a separate section in the second floor, which has a good ventilation as well as an appropriate view to the periphery; moreover, they are away from movement and activity centers.

A four-season room that acts as a greenhouse is in the second floor and southward. It can be used as a living room and a small kitchen for family members. When designing the ambit I tried to use low-height vegetation to create a pause space among the condense periphery trees. However, in north and west directions high vegetation has been used due to adverse winds.

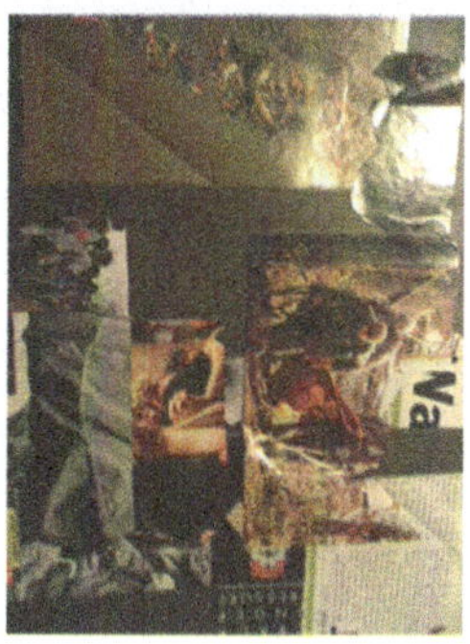

3.14 The Hidden Identity

Designing architect's dwelling

Student	Shila Ahsaee
Masters	Dr Ali Asghar Adibi
	Dr Ahmad Ali Farzin

Awakening the slept creativity

There are various ways of obtaining a strong design and effective use of mind creativity, which we experienced one of them in this semester called "designing according to the collage". We cut some pieces or pictures of newspapers, magazines, and books with regard to their color, form, and subject, and freely arranged them according to light, color, space, and material and thus obtained various collages using personal taste in which the presence of light, darkness, surface difference, and color harmony can be observed. The ideal space of these collages in fact showed us the spirit of the work, and accidently introduced our ideal and interesting space for living, resting, etc. after obtaining various collages and consulting with the masters, we extracted this ideal and interesting space from the collages. We installed on the wall the collages in a distance away from where we sat and extracted these spaces by focusing on them and sketched them. The result was various sketches with a high compatibility with our spirit and each one was more complete than the previous one. This shows the expansion of our view and greater success in extracting ideal spaces from collages.

Creating an ideal space

Then, a maquette was created according to sketches and collages that should indicates all details of the interior space of the element idea in which colors and genders should fully represent themselves. Collages are still sources of inspiration. The color and light harmony were introduced into the work to display the real identity of collages in the maquette.

Thus, we arrived at a maquette of interior space called our favorite space. In this stage, we should consider developing this space and reaching a complete, ideal design.

I had problem communicating these collages at first and only some repetitive designs destitute of identity existed in my mind, which I forced myself to use them.

Because this was my first experience in collage, I dealt with it quite primarily, but in later designs, I will be more successful through nurturing my inner creativity.

After designing the plan, I started designing the façade and then created the final maquette and thus the task of our designing an architect's house ended.

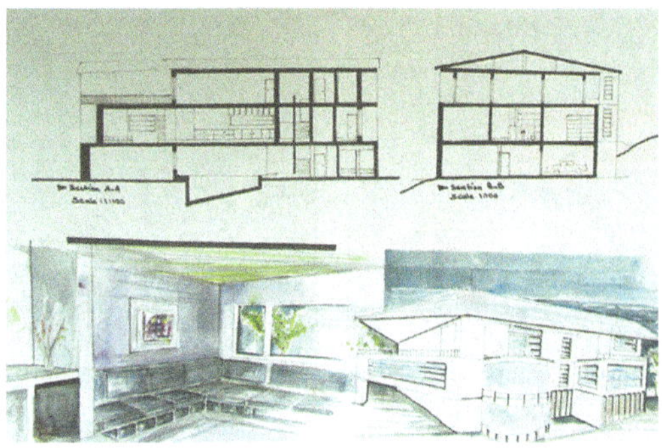

Above: final maquette
Below: collages and sketches

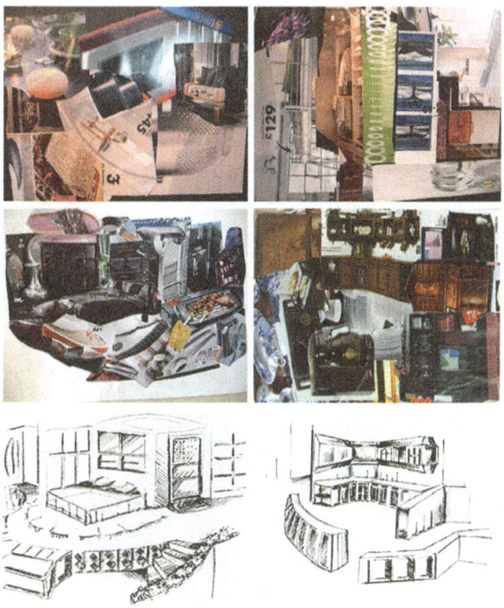

3.15 Awakening Pictures in the Memory

Designing architect's dwelling

Student	Fatemeh Hashemi
Masters	Dr Ali Asghar Adibi
	Dr Ahmad Ali Farzin

The human being and his life environment are of a complex essence. Moreover, there are numerous issues having different characteristics. The need for new methods and making the pictures visible open up an extensive field for creativity and developing imagination power and these methods should be exploited.

One of the educational goals of this semester was to nurture creativity potential. The ease of designing and dealing with new issues requires speed, accuracy, and being attracted by unknowns and unbeknown. It also requires risk-taking and search spirit and being released from rigid and formal contracts. Given the issue of design 2, i.e. "designing a house for the future", any individual needs to create interior spaces with regard to his moods. To achieve this goal, visual tools method was used.

Given the issue of design 2, i.e. "designing a house for the future", any individual needs to create interior spaces with regard to his moods. To achieve this goal, visual tools method was used.

The "picture" in our architecture shows a tool we think a tangible reality can be introduced using it. To achieve these goals, it is important that the result can be transferred to others. When we stand in front of these pictures, this visual tool evolves us and creates actions and reactions. The mind is activated and, by referring to the memory, the picture of various spaces from the unconscious mind and innate predisposition is extracted.

The similarity of spaces in the memory are analyzed and criticized. A relationship is created between the spaces, are finally turned into a metaphor, and emerge in a tangible form.

Awakening pictures in the memory, understanding the similarities, turning pictures into a metaphor, all of them nurture the mind and creativity.

I drew sketches of them. In these sketches, the surface, light sources, and the general space essence are introduced. Based on the sketches, I selected the collage for the rest of my work having a shadow of the peripheral spaces in it and it could be situated in the heart of the house. I considered this collage in more details. However, other pictures were used to create the details of the design.

Above: main collage
Right: collage and the sketch of the collage

Above: final maquette
Below: maquette elements

Analyzing the collage
By looking at the pictures infused by colors and material, I felt being in a warm family sphere. Thus, undoubtedly, this picture does not imply a private space like that of a sleeping room and introduces itself such as the private living room of family members. I analyzed collage elements part by part to create peripheral spaces and developing this space. For example, there is a light reflection in the left-hand picture, illuminating part of the short wall in front of us. I defined an opening to the patio based on the existence of this narrow light, which acts as the main entrance pattern. There is a penumbra in the end of the space, between two end walls. My interpretation of the space between these two walls is a staircase toward sleeping rooms. Based on these examinations, I developed the primary plan obtained from the collage and the final plan was designed through organizing the plans. For corresponding the plan with the site regarding roadway access, pedestrian access, view and scope, ground slope, etc. I compared it with north-south axis symmetrically. The final plan corresponds with the site and its characteristics in all aspects.

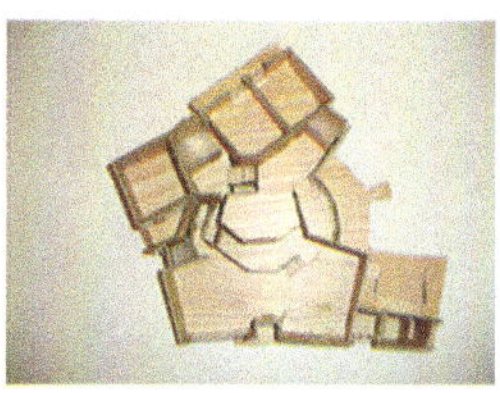

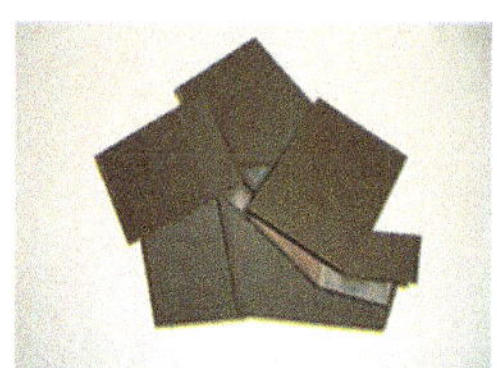

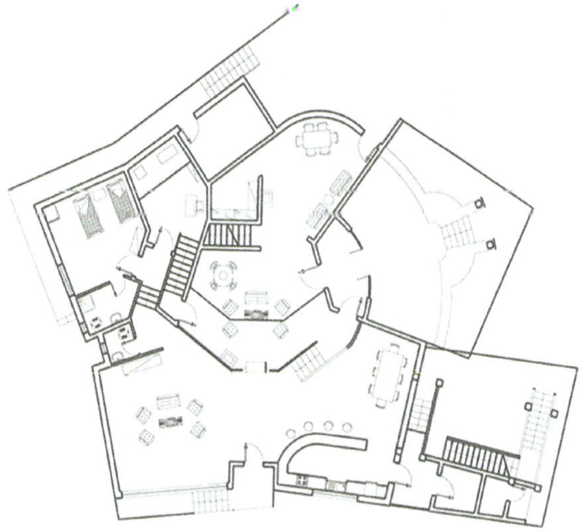

Above, right: ground plan
Above, left: project sections
Below, right: first floor plan
Below, left: perspective of the project

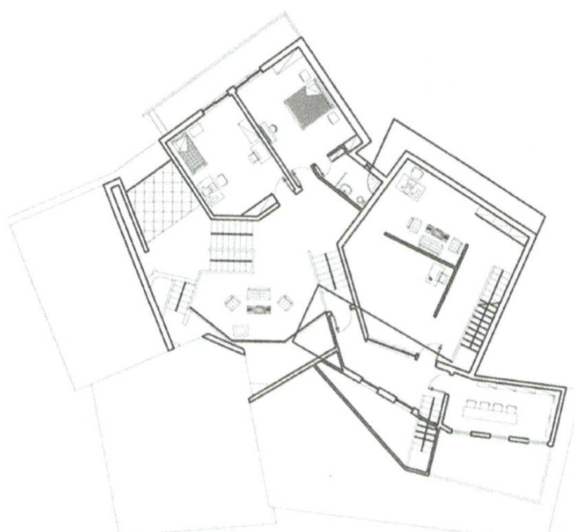

3.16 From Dynamism to Inertia, Without Cuts!

Designing architect's dwelling

Student	Ali Mansouri
Masters	Dr Ali Asghar Adibi
	Dr Ahmad Ali Farzin

Second collage

Third collage

In any enterprise, work is the start of the road. The same applies in designing. The thing we call designing process is in fact a set of choices illuminating the road to achieve the final result. In the process of this design, we confront a new option; designing using the collage. The results of this method were incredible in recent years. A set of intensely 2-dimensional pictures had created real and tangible spaces and when I looked more carefully, spaces had an evident and hidden harmony with early pictures. Despite encouraging and repeated comments of the masters, I opted the road of collage with a sense of ambiguity and adventure.

As the route of this designing method has been defined in details in the introductory of this book and comments of other friends, I avoid repeating the design process.

The second and third collages

Due to reasons to be explained, I start from the second and third collages. The collages in the first glance I believed had no deep relationship with me. I accede that when arranging these collages, I tried to fit together those pictures capable of producing space. This tendency was lower in the second collage and culminated in the third collage. However, the result was disappointing. This was the reason these two collages and all their lights and colors were forgotten for a long time. I put aside the third collage forever, but the second collage came in handy later.

The first collage

In fact, the first collage I "authored" was the most important collage. My opinion was the same as my master that given the time available to me, this collage was excessively abstract and demanded interpretation. In first glance, none of the elements of the picture attracts the attention to a particular space, especially that big and live knob insisting on being "the door knob" even in the farthest distance to the collage! However, the colors and lights of the collage were interesting to me. Thus, I concentrated, looked, and drew, and finally arrived at an acceptable picture. A smooth space with mild lights. In fact, the obtained sketch had become more complex than the collage itself. For example, my interpretation of the powerful presence of that knob was that such a tranquility and concentration exists that doorknobs, even in a long distance, greatly influence the viewer! Alternatively, I had placed something in the semi-translucent, middle wall that even myself did not knew its meaning; a window to be opened on darkness! (Moreover, later put it aside for the lack of enough time). Even the strange roof that huge knob had fallen on the space remained unknown for a while.

However, in fact, the first and second sketches were depicted simultaneously, which were extracted from the left and bottom sides of the collage: a space greatly relying on warm colors of this part of the collage, and with a new idea: the pendant, wooden bridge passing through darkness, submerging in end of the light. Intense dynamism of surfaces, warm colors, artificial lights of spaces, and curved staircase were among the characteristics of this sketch, protruding from the orange stains of the corner of the picture that had penetrated into upstairs.

A maquette was created based on the sketches of the first collage in which the lights of spaces and general spirit of the collage were emphasized.

Pictures, from top to down
First collage
First sketch and the plan
Second sketch and the plan
Pictures of first collage maquette

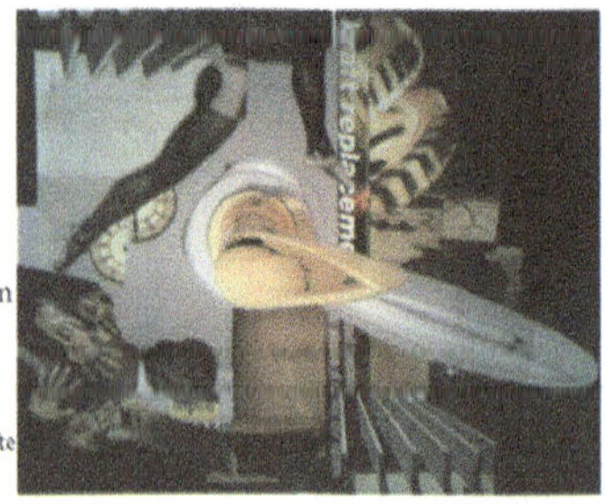

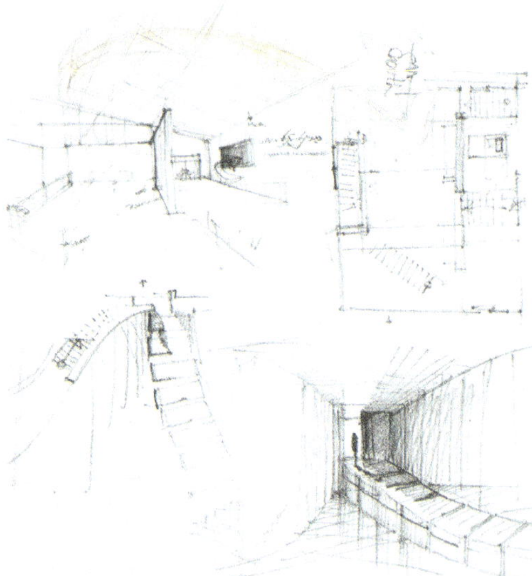

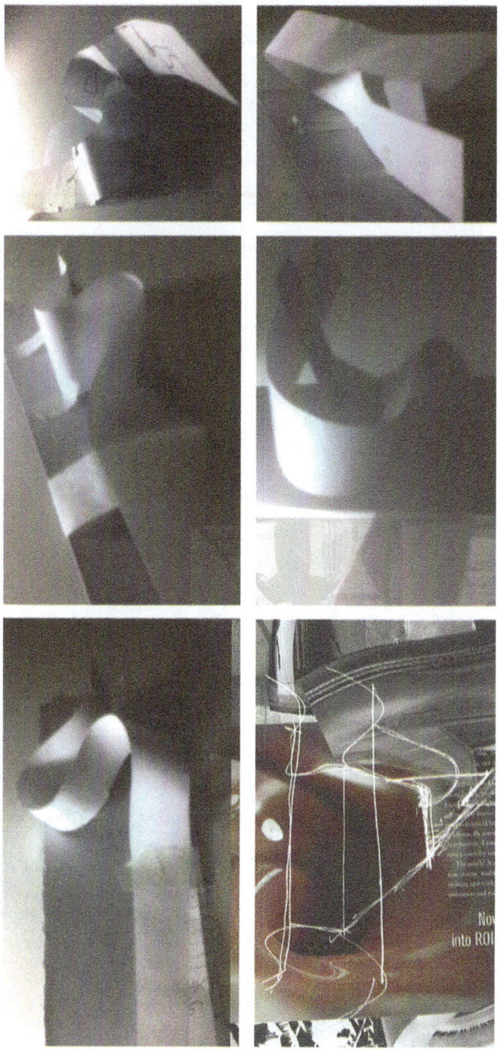

Two depicted sketches were highly contradictory;

One side, a bright and flat space and the other, a dynamic and dark one. In doing so, these spaces were improved as two level structure (slope of site makes it possible). Therefore, my house was defined as:

> Dynamism of exterior space extend into more public and active part of interior space. While we are going up to spiral stair, its curve is gradually reduced, till we get to silence and calmness of private space.

We confront mentioned definition in two ways: First, all events, from public to private space, are only obvious in plan. A plan, full of curves, for ground level, and linear one for upper level. Second, I represent space modifications in movement of one or two continuous planes; A plane, in manner of continuously rotating wall, which start to rotate from ground level (its rotation form a stunning spiral stair) and end in upper level and provide a ceiling or floor. There was also same rotation in second collage and it wonder me and if I could imagine better, I might achieve these rotating stair in initial collages. A stair, in a free planar cover, rotated and ascended. This rotation needed continuous movement, from vertical to horizontal, of membrane without disturbing the stair.

To solve this problem, It required different continuous paper planes which I designed them differently to provide variety ways and stairs, in the middle, to reach upper levels.

Actually selection, in this part, was based on aesthetic, function, proportion of space occupying and selected plane. Beside irregularity of plane which had not specific geometry, it should not exceed more than usual in order to not only preserving dignity of structure but also drawing of planes were improved to design a plan.

Before final selection of plane, I had to draw a plan based on site analyzing and studying of personalities of people which planes were defined inside it later. This scaled diagram of correlation, after planes locating, was modified due to some reasons. First, scale of spiral stair was almost huge. Second, it was view to stair in order to make visible reference point of structure. Moreover, after locating planes in volume, it was time to define axis of column which it also caused modification.

In this step, final function of planes was defined which both were changed in sloping roofs of ground and first levels. In addition, plane, moved beside bridge and went down, formed an entrance when it was going to be horizontal. It went out horizontally and created a semi-open space. Moreover, this plane along with a plane which appeared in the middle of north wall were considered as connection points with exterior.

Following step was formation of details including color, light and furniture base on collage. For instance, the wooden bridge at the top of empty space and fireplace of living room. Parking was provided by eaves of first floor. It made possible to design a wall in location of column. Extension which formed by turbulence motion of two planes, provided entrance. Angle between two plane in interior caused view and motion an manner that view to private space was almost limited and motion unconsciously directed to living and activity rooms. The matter which leaded me from beginning, is that spaces where social life and activity happen, there should not be inaction point and home is motive of life. Moreover, when families need to be alone, home as a membrane keeps them away; in continuous path and no pause.

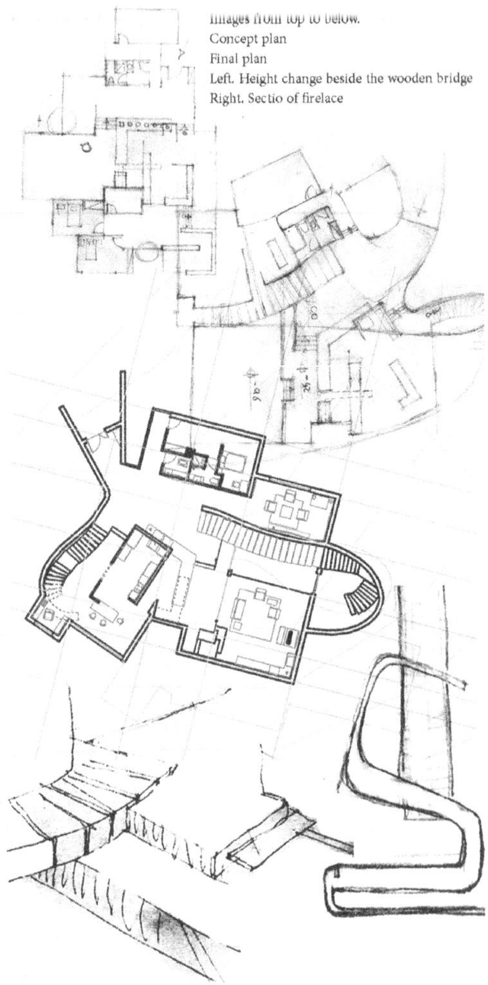

Images from top to below.
Concept plan
Final plan
Left. Height change beside the wooden bridge
Right. Sectio of firelace

Then, we continue the project to make our ideal images by collage technique, sketch the result images and make a model of that images during two weeks and the goal is to arouse our hidden mind and find out the general concept.

In the next step, not only the special point for instance the color, light, material, sound, interior and exterior relation, space dimension and proportions etc. but also their positive and negative points will be reconsidered.

Finally, knowing about intended standards and surfaces, we select the most suitable physical and psychological dimension, moreover, regarding to the site and choosing the best place for cultural center, we study to find the acceptable location in terms of accessibilities.

Characteristic of the cultural center

The Chizar cultural center in local scale, regarding to the space for service, education and leisure

1. Amphitheater for 200 person
2. 5 educational-workshop class for 30 person
3. A 400 m^2 exhibition hall
4. Library for 3000 book
5. Space for office work (management, 2 official room, service, etc.)
6. Other space (buffet, praying room, entrance place, shop for selling products, installation, rest rooms, storages, etc.).

Design site

Design site is in Tehran, that is shown in the in following page figure.

3.17 Chizar Cultural Center ■

Introduction

Designing is result of subjective and objective activity, needs and realization of project special functions that is hidden complicated step of creativity and logical action in it.

Designing is kind of process that is called «design process», but sparking at the beginning of formation is necessary. If this lightening getting up from designer, it will be valuable.

Before starting to design, we should think about subject and analyze the detail and basis of project. The aim is to deal with the subject, analyzing the work sample and site knowledgeably.

Aim

The project purpose is to let student know about interior space of an architecture structure, that in this the effects of space interior function to space exterior form is being considered. These consideration is being dealt with in three below aspects:

1. Realization and analyzing interior space function and consider their regulation and relation of these function;
2. Selection of Material, element and detail of interior space based on their importance and suitable composition;
3. Pay attention to the space elements effect on each other.

Design steps

At the beginning, in addition to visit the project site and analyze the case studies, we start to sketch then study the effect of different environmental factors on the existing design and evaluate the function and their relation to understand the elements and whole the project entirely.

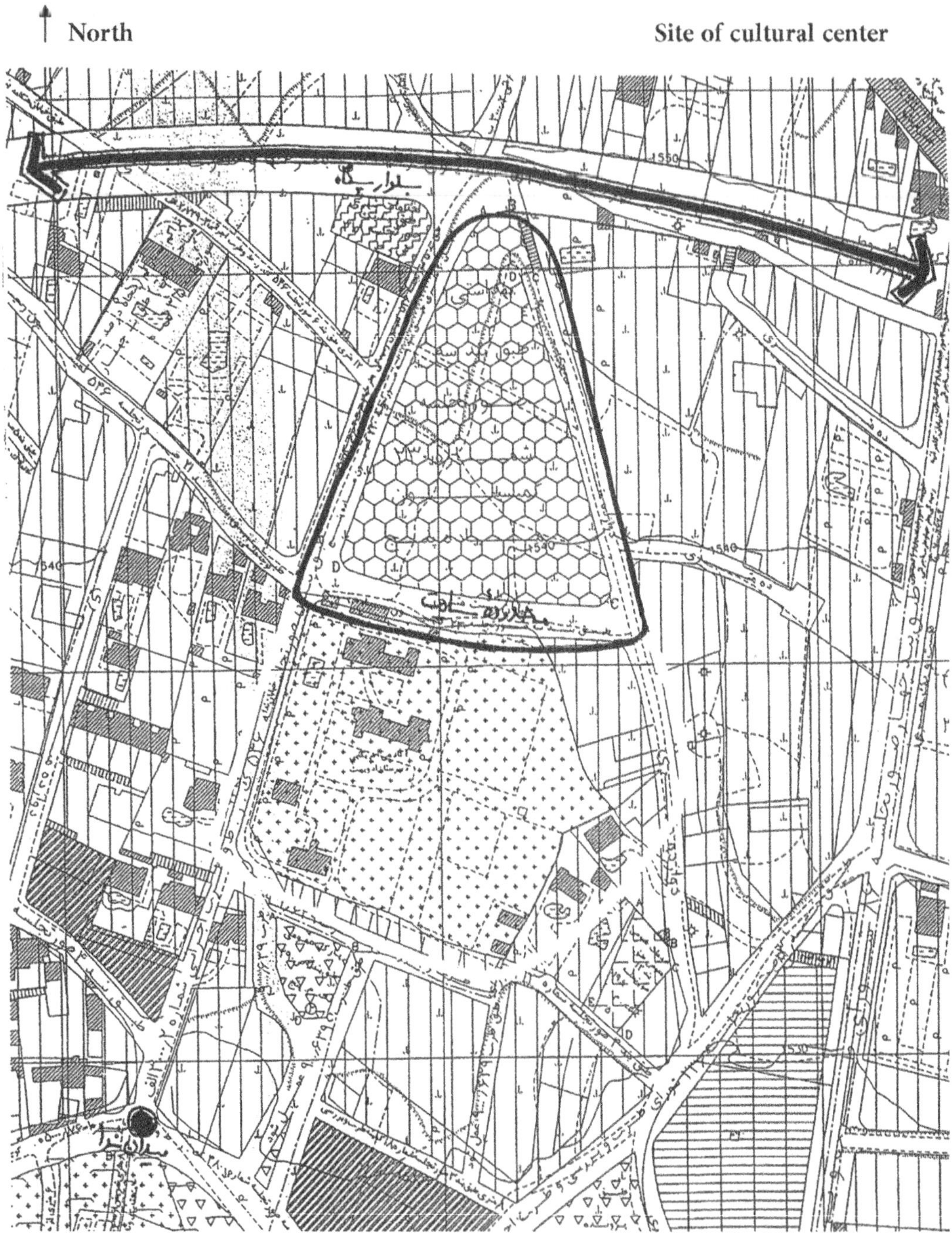
North
Site of cultural center

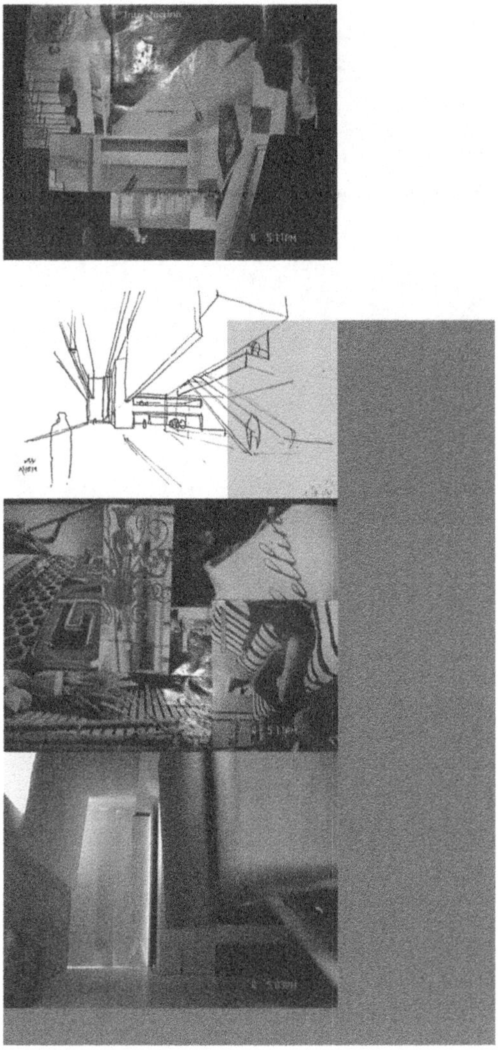

3.18 Cube

Designing the Chizar cultural center

Student	Babak Shariati
Masters	Dr Ali Asghar Adibi
	Dr Ahmad Ali Farzin

Third architectural design was the second experience in the collage group. I think explaining collage technique is not necessary but to explain summarily this technique, the newspapers or journals clippings assemble near each other in abstract way to create a new whole. The portions is selected by your enthusiasm and the result images will be formed considering nothing. Having intended space collage, designers are able to sketch the space from the collage and then make model from the sketch. The interior model give us the space which is definable or even developable, after the space is developed the design space will be created.

Experienced the collage with the colorful portions and too many varieties, I decided to use less variety and more balanced colors, so I respect my eagerness and pay more attention when choosing the images.

The two first collage, unlike the third and fourth, were kind of warm-up and the third one was the main, so I sketch from that to make a model.

Although the image were few, they glued close to each other because of big images. The color ranged from light cream to dark brown.

The collage space seemed shiny, deep and high. Materials were burned wood, concrete and stucco and floor sounds glossy. I made and effort to make a model in accordance to the sketch and the sketch lightening perceive on it the same.

After modeling the interior space, the fol lowing step is studying and analyzing the spaces, their relations and summary of analyzing works. I divided spaces in two categories: the educational space and other. The basis of category was silent, so the class and library were in distinct category from other spaces.

Spaces including restaurant and coffee shop, is separated and the reason is this space are open even when are close.

As a result, it is better to locate lobby, galley, Amphitheater and official spaces close.

Accessibility and entrance were considered in site analyzing. Eastern and western Street were narrow but the southern street is wide and easy access, so it has capability to locate the main entrance. Except the school in southern street, whole around are 3–6 floor residential buildings.

After this kind of analyzing, it was time to come back to the collage process and define functions for model space. Lobby was the function that I make a choice for it, in this way, I find in that model, the place for galleries, amphitheater and offices. The space density project forward the architectural mass to the cubic form that I consider this problem to mass imagination.

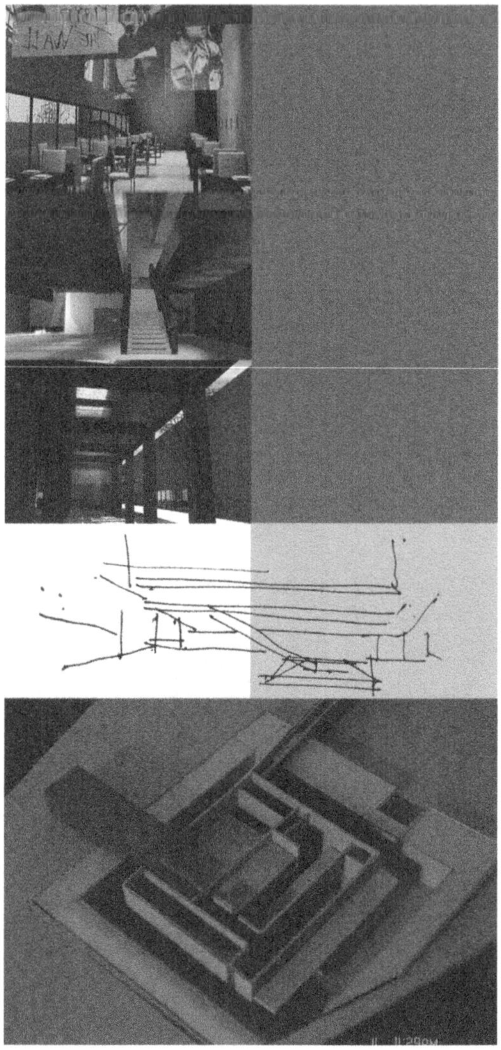

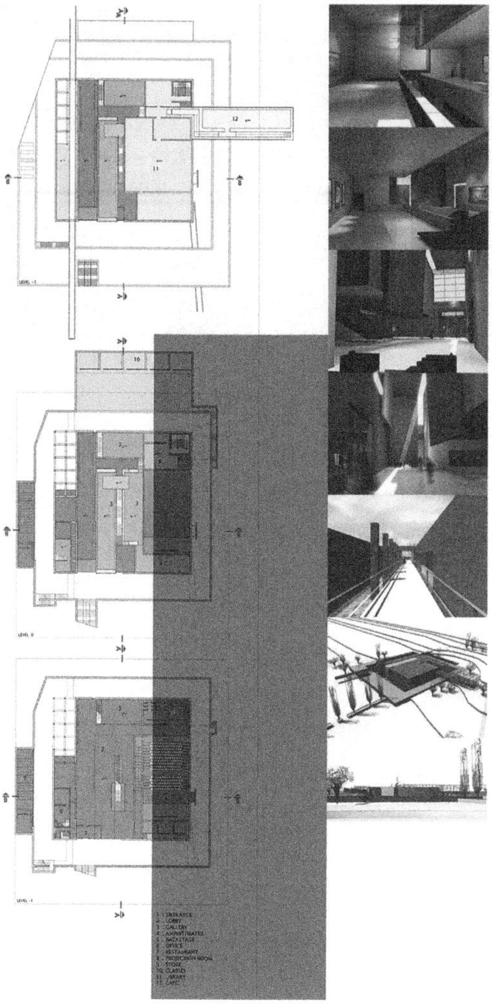

I started to prepare the plan and arrange the spaces same as model. Except at least changing in dimensions and areas, no changing was necessary. In order to control the whole mass, I designed the plan in square-shaped and place lobby, amphitheater, gallery, official part and rest room.

The heavy hanging mass seen in collage and model, functions as gallery that the visible staircase the access path. The galleries have connection to form an integrated gallery in great exhibition.

Lobby 12 m high in its main parts and a ribbon of light near the ceiling guide the eye toward the space depth. Behind the hanging surface in south of lobby was a way to penetrate light. The west part of lobby should be made empty to have this light into space that in order to have a good obligatory cubic mass, hypostyle space in ground floor, pathway in upper level, it was created. Library located at the top of amphitheater and planned an access way from interior.

Now the result mass is cuboid form with square-shaped section and 12 m high which its height damage the surrounding buildings view. So I decided to deep the building down into ground and surround it by ribbon of yard. The barren yard is 8 m lower than the ground and the planted ground is the sole green landscape that is observable from yard to ground.

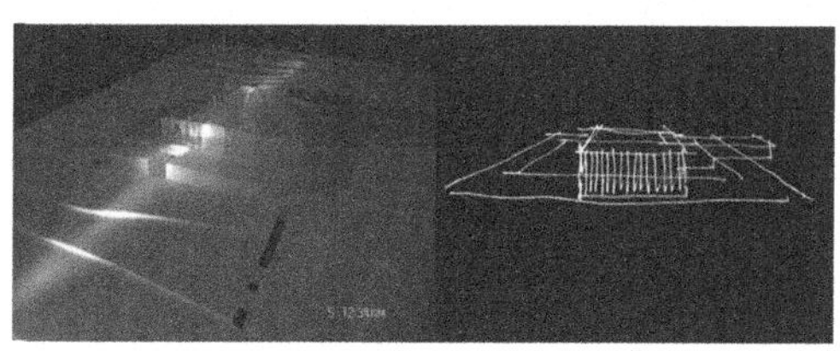

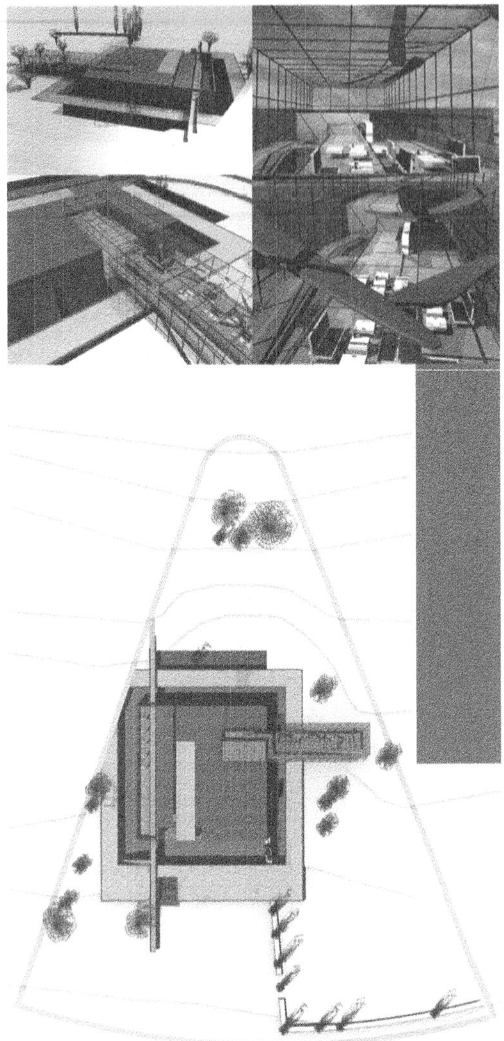

Building a pool in the upper ground around the yard was another idea that passed by, to reflect the sky and mass and also show the mass hanging. I also could place the restaurant under this ribbon of water.

The other structure that should be away from other building was coffee shop. Since coffee shop same as restaurant should be opened all days and nights—even holidays-separate access way is necessary. So I decided to design the coffee shop as intermediate mass between ground and heavy mass. The cubic and glass material coffee shop start from ground and penetrate into heavy mass. Being light mass, there was contradiction between coffee shop and heavy mass. From distance, coffee shop seems a hanging plane that people walk on this horizontal plane. The exact location of whole mass was known based on the main entrance from the lower part of site, parking place and present Populus trees.

Have a courtyard, a ribbon of pool and parking, there was no more place remained.

All designing step had been done through free hand sketching, model making and primary plans. After these steps, the final design was modeled in computer with somehow changing. At last, I compare the rendered image of lobby by the collage, sketches and model of interior space.

In conclusion, I experienced in second time, it is possible to design interior space light in the best manner by the image of a excellent composition. So finally we are able to design a space light base on the rendered image light. I think that this class was more successful than the first time.

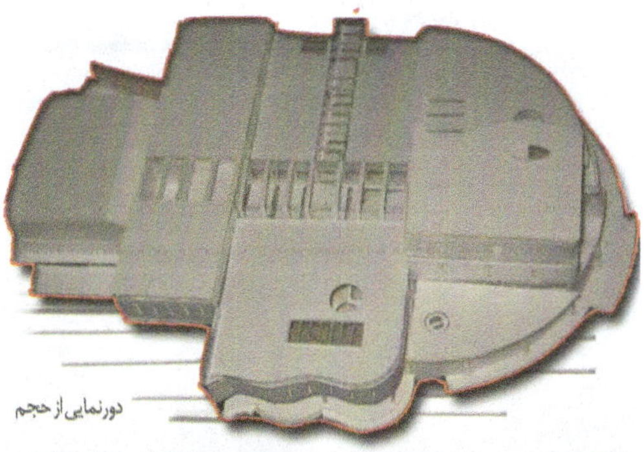

3.19 Creation of Fluidity in Space

Chizar cultural center
Student Yasaman Haji Esmaili
Masters Dr Ali Asghar Adibi
Dr Ahmad Ali Farzin

Again Collage, experience of excellent time and thought flying toward unknown world. This time, collage is being served to create a public place, design a space to spend beautiful moment.

A place to walk around and discover creative idea in each side. Spaces for children to have connection with each other; Place far from busy city, car pollution and to experience new world.

Transparent wall for the one who escape from the repetitive life and gather together and spend their wonderful hour of their life.

Flexible planes slide on each other and people can walk around, sit in pleasant place and look around.

Time spend and people because of in their busy life accompany with each other and move forward. The only reason is willingness of people.

They are part of my thought to design a cultural center according to people interest and their needs; so I select the photos and paste on a paper to create my ideal space.

The first collage

The first chosen photo is reference point of my space. The point that start and end everything from it.

A crooked cross. Soft light make bold this cross that seems I can touch boldness.

Spaces have been design along main axis. Collage space is full of people. People are not only representative of themselves but also representative of space and dominant feeling on it.

The second collage

Movement toward light, full of freshness, in narrow space that become wider at the end.

Trees and plants grow up from the lowest part of space till they blossom up on the floating surfaces of the chess pieces.

Points that all part of space move toward them are behind the forms and shapes. To create them in our thought, they are invisible in our view.

Shiny floor reflects the sky light and make the space bright, fresh and also secure.

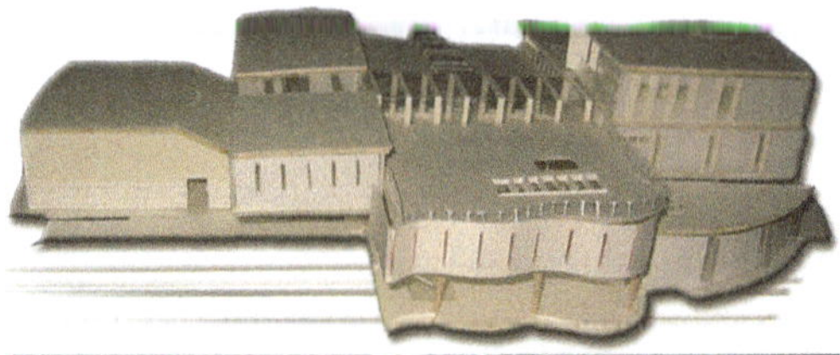

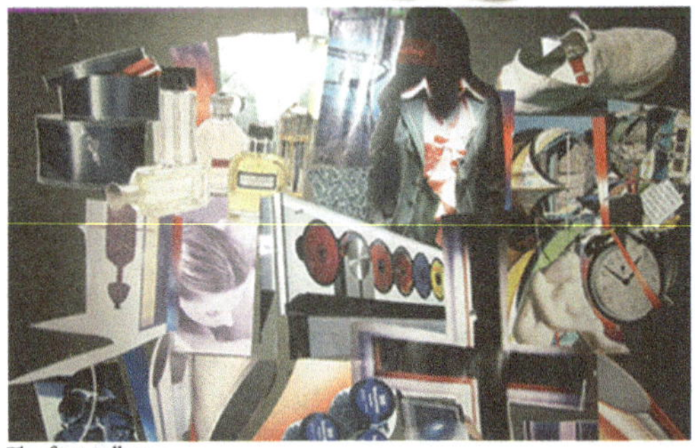

The first collage

The second collage

The third collage

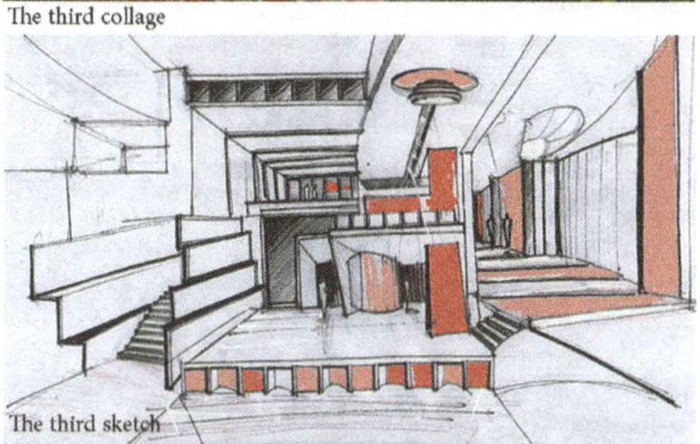

The third sketch

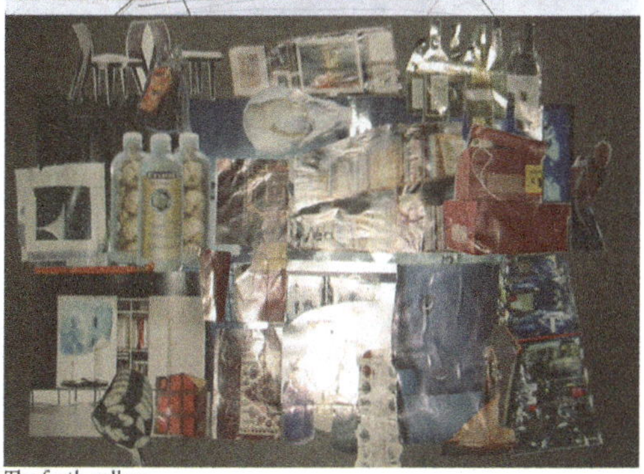

The forth collage

The third collage

Human being come out of access point and move, dance or run into space. There is significant correlation between this and previous collage. Part of space along with its terrace seems hanging. People can observe the stellar sky.

Colorful and shiny mass blossom in spiral movement from this space. Space had been designed like mezzanine up on each other that rotating stairs connect them. A hole in the ceiling covered by convex glass make possible to observe the blue sky and because of glass characteristic the observed part of sky seems more extensive than usual. The central hall, with the purple and reflective floor and mirror decoration walls, is more deeper than meets the eye. The corridor on the left side of the hall, stand firm on the ground, invite us to get into it and discover other world of cultural center.

The forth collage

This collage function as variety, small and big spaces, for instance a library and classes. Like the previous collage, here again there is a background of night sky. No different between the color composition of this and other collages, sparkle of red light in blue and purple background scene along with bright light like sun shine. I feel that cultural activities will be done rapidly in spaces created in this collage. Book protect the history narrative and secure the culture. In addition, Iranian carpet represent the my country traditions.

The forth collage
Life, the fresh life, can be seen in all thing horizon. Children movement on muddy domes and blue sky full of transparent cloud.

The forth collage

Red planes flow into water surface in front our eye. Muddy wall stand firm at the back.

In each side, there is new discoverable space, where we can play, walk around, run, or relax, draw, stitch a carpet or make pottery.

Actually, this collage show landscape and mass connection. This initiate idea help us to design site.

Finished the collage, model have been made based on abstraction in collage.

کلاژ، کروکی و ماکت حجمی از فضای رستوران

Thus restaurant was created. Free movement of main wall cause a fresh space and the plants connect the planes at the center.

Passing by the rotating stair, Music sound at the second floor excite us. Maybe we stand at half way or climb up small stair to listen the music at the second floor.

At the end of model an opening toward sky is defined. Blue sky and fountain at the first floor catch our eye.

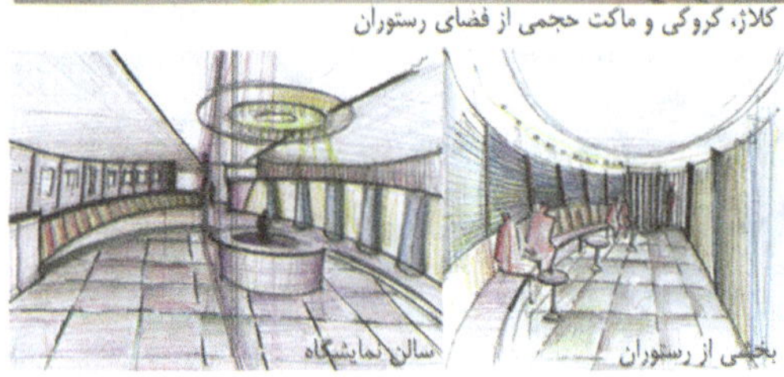

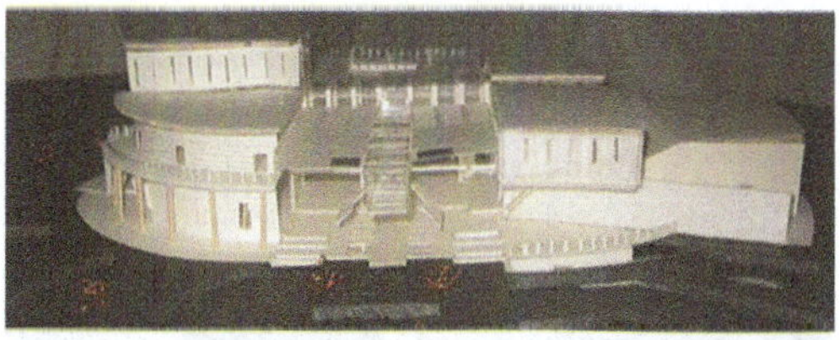

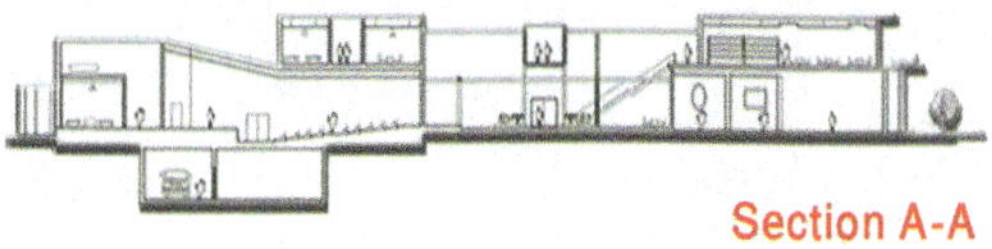

Section A-A

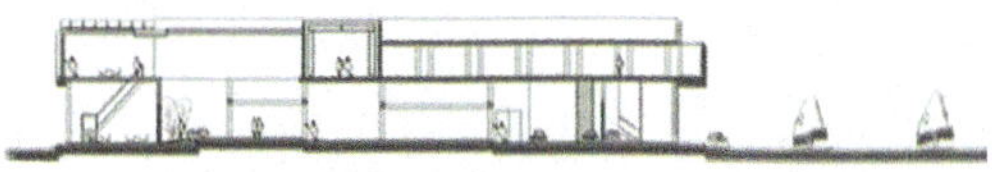

Section B-B

East elevation

South elevation

North elevation

By composing the collages, spaces is defined and all together create the cultural center.

A central glass corridor, visible of surrounding, make possible connection among spaces. Beside entrance, mezzanines for lobby, restaurant and exhibition are defined.

Restaurant as can be seen in collage is designed so that we can observe surrounding landscape from the first floor and interior space of cultural center and music hall from the second floor.

In exhibition as the first collage, glass volumes come out of one side of it and reflect red light to objects, similarly, flexible forms are defined on the wall of other side so that rotating on its axis, landscape and variety status of light can be seen in exhibition.

Amphitheater, the other side, stand firm and its order and symmetry induce peace and silent feeling.

In second floor, classes and library on one side and official room and computer site on other side are defined. In third, there are two workshops as governing space on all site.

Finally, direct and indirect access path are defined to among spaces and also to exterior space in best way.

Soft lights in ceiling are designed so that when variety type of light penetrate into space, their reflection on the wall and floor cause diversity in space.

Flexible space is also firm and silent in order to induce peace feeling. The crossed form mass connect all space together. Places where walls and ceiling or decorative beams and column join, this form is obvious.

Form continuity in surrounding landscape is also obvious. In landscape design, some places to sit and relax are defined.

Artificial lighting is designed into fountain to create a wonderful space during nighttime hours. Lighting is used to emphasize on the main entrance and also contradiction between void or occupied spaces. In addition to that, lighting cause beautiful parts of mass seem more wonderful.

In each side, light invite us and define our movement way in site and ensure us where to go. In other word, lighting is our guidance and also forms space and volume.

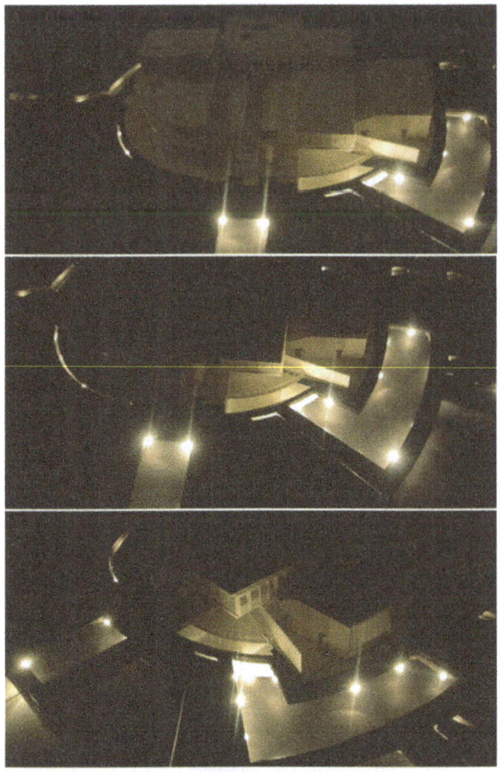

Night views of building and its surrounding site

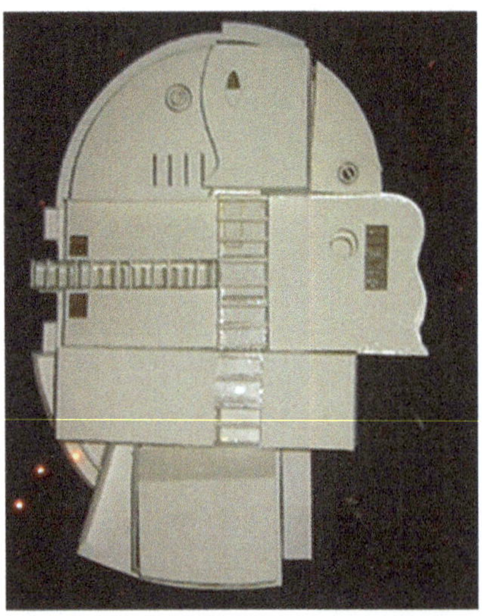

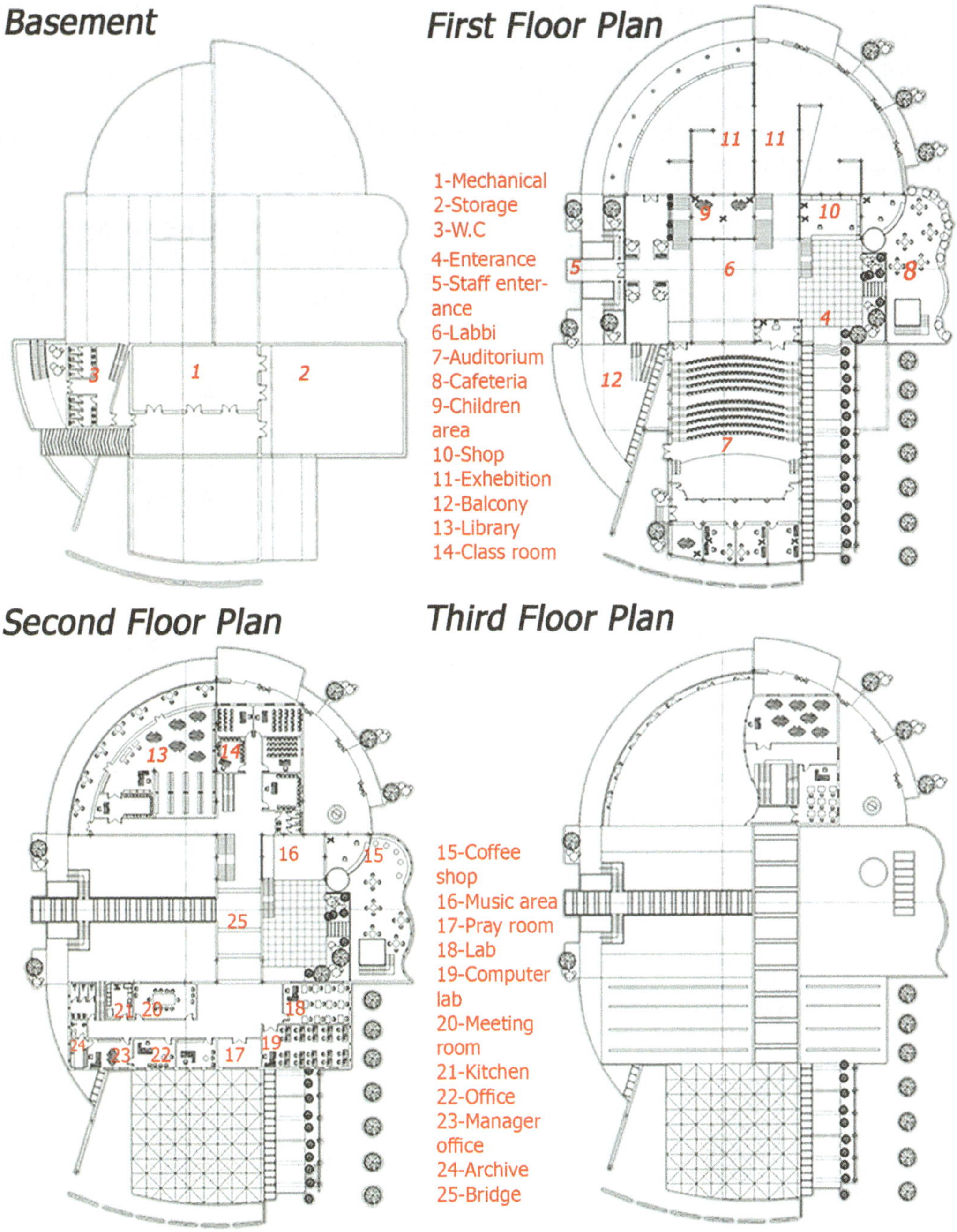
Basement
First Floor Plan
1-Mechanical
2-Storage
3-W.C
4-Enterance
5-Staff enter-
ance
6-Labbi
7-Auditorium
8-Cafeteria
9-Children
area
10-Shop
11-Exhebition
12-Balcony
13-Library
14-Class room
Second Floor Plan
Third Floor Plan
15-Coffee
shop
16-Music area
17-Pray room
18-Lab
19-Computer
lab
20-Meeting
room
21-Kitchen
22-Office
23-Manager
office
24-Archive
25-Bridge

Based on main idea of design and using the site collage, the site plan was designed.

A mall in northeast and a wooden cottage for art class at the top of the hill in northwest of site are considered. To have a beautiful landscape for exhibition, a plant fair was defined.

There is a parking and a amphitheater in southeast, next to main entrance.

I use collage in whole design steps, from the first spark of design till end, from main idea of design till completing the mass and architecture design.

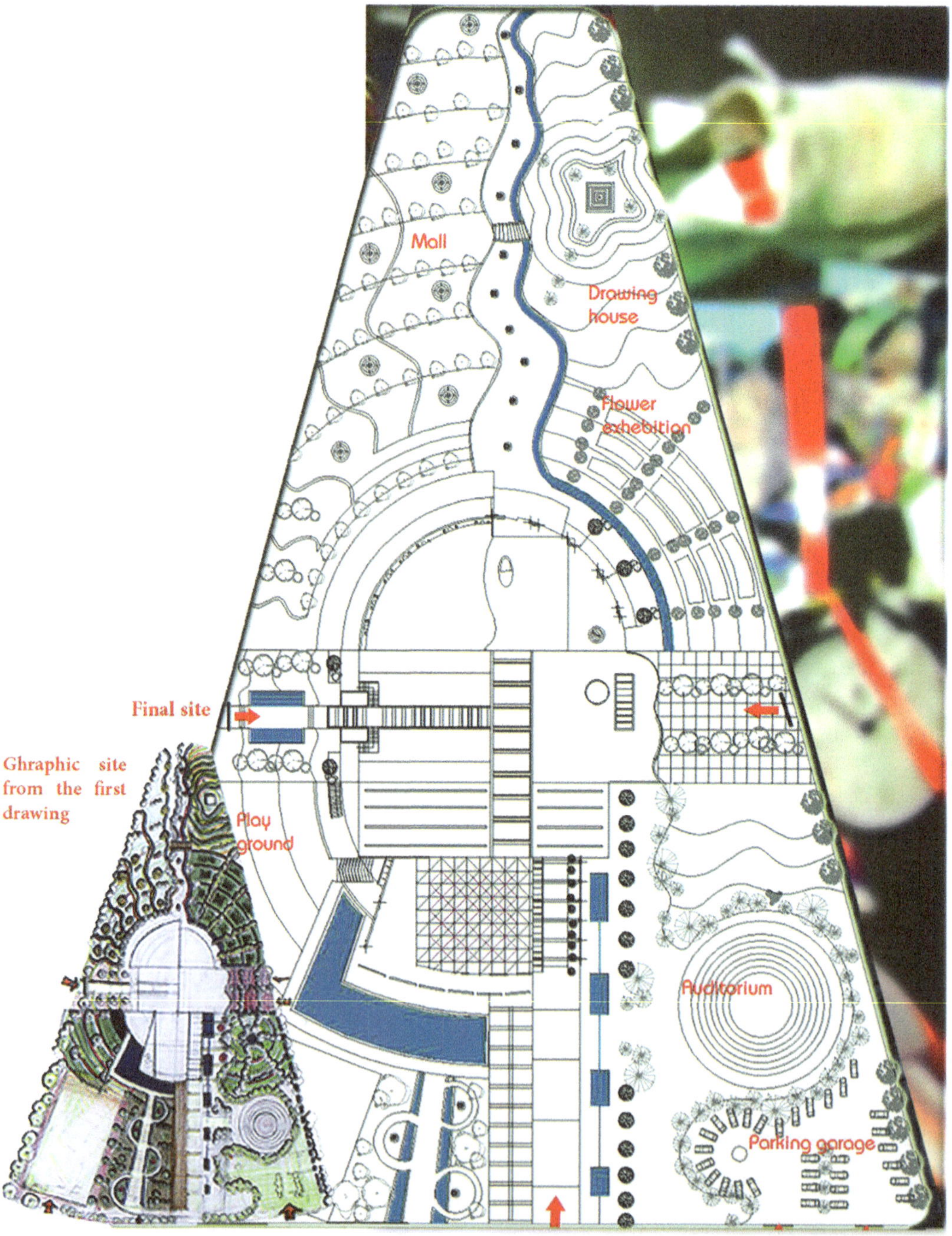

Arasbaran cultural center

The design site

3.20 High, Light, Transparent

Chizar cultural center
Student Anahita Khodadadi
Masters Dr Ali Asghar Adibi
Dr Ahmad Ali Farzin

Designing the cultural center was began by analyzing the characteristic of the present cultural center including Daneshjo and Arasbaran cultural center. we visited and took photography of the mentioned Arasbaran cultural center, when people of Abbas abad region were preparing for the days of Tasu'a and Ashura. Some people were practicing the Tazi'eh and a narrator told a epic of Shahnameh and Ashura in entrance lobby of cultural center and some came to the class there. An artwork exhibition about battle of Karbala and Shahnameh was organized. Generally, groups of people were doing activities in each side of cultural center in order to hold the ceremony as well.

Chizar site analyzing including weather, culture, people behavior and building usage made it obvious to select my desired images in my thought that was suitable for mentioned space.

I chose my desired images and it was time illustrate my idea in by collage.

The first collage
When I was looking for images, Lightness and vastness of space was the first thing attracted me. Glass and steel made network plane induce feeling of lightness and vastness. The space was like a natural landscape of dessert, with sky and white lines on it, it not only cause heaviness but also give variety. Beside the calm blue space, dark red bricks caught my eyes. I like the lightness continue through the arch above the space. Second piece was chosen network plane, which its color was same as red brick of the last part of first collage.

It was better to pay attention to the reference point of image, the redness of brick at the end of space, and use this attraction in collage.

Curvilinear line pattern of floor represented a wavy path that ended to a light space in a red colored context. The question passed by in my thought was: what happen out of this space, not far from it. Maybe there was a porch around the building or a central house and planting site at the center.

The second collage

Intensive light, penetrated into space from ceiling and cause along the spiral form stair, ramp or something like that, was chosen as main factor in collage. I extended the spiral movement around light corridor and enlarged the downstairs ceiling parallel to other curve toward infinite point. It provided a vast and friendly space to public gathering and watching theater, movie and music playing. Space, down the ramp, should be a cozy and comfortable and also warm and bright to speak to your friend or stay there for a short time. I chose dark red and white to emphasize on red color of context. To avoid a massive space and find a way toward or provide fresh air, a garden was defined behind the wall that remind us plant blossom in the spring.

The third collage

I preferred to walk out and notice the exterior space. There were some frame outside of building but I could not guess What they could have been. I colored the horizontal and vertical elements of frame white in order to set up desirable contradiction between frame and dark green background of out. The first one, white vertical element with a rotating cap up on it, defined a place to stop and rest that was lower that the ground surface. Sun shone on the long beam of porch, the porch around the first element, and caused dark-light on the building face. A stone stair went up and cut the porch and ended to a green landscape. On the other hand, the reflection of tree on the blue plane around the porch that wind cause it unorganized, could be seen on the floor of space easily.

I sketched the resulted colleges. The black background collages were my vague and general image of ideas about considered space. In this step, I added or subtract to this image whatever was necessary.

The second collage that defined rising spiral space and porch around it and open light space to speak and rest at center was chosen as main mass of cultural center. First, I made a amateur model based on the sketch of second collage. Long bright hole and dark-light caused by the porch on the ground were defined by slice plane. But I had to show the bright state as best as possible in the next steps.

The next step, I considered the model as a real project and upgraded it. The first subject that occupied my mine was, how it was possible to keep the high porch as hanging planes, and the second was, which part of building the light behind the rotating stair should penetrate into interior space. I considered and analyzed different ways to solve them. The result was a mass. passing by its entrance and light corridor, there was a spiral stair. Upstairs, a space to public activity and gathering was defined that some part of it continued through a hanging porch that in a spiral rotation ended to the ground.

Model of collage

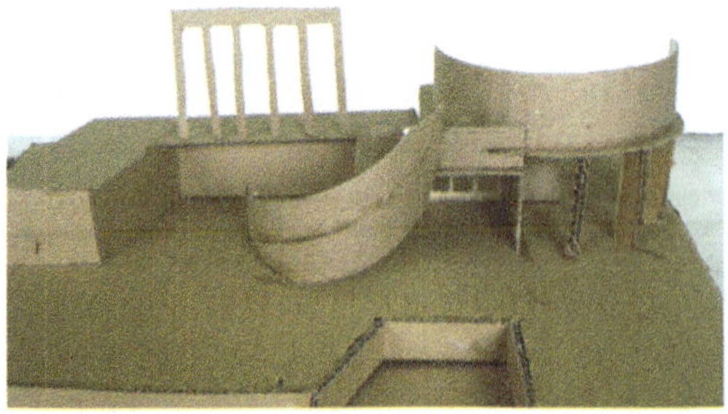

The first model

The interior sketch

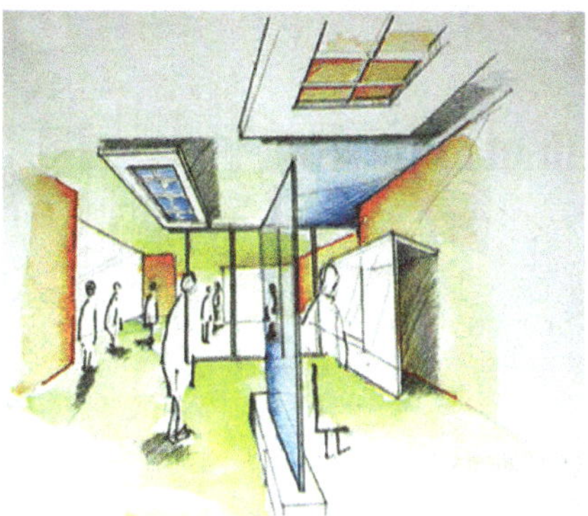

Gallery

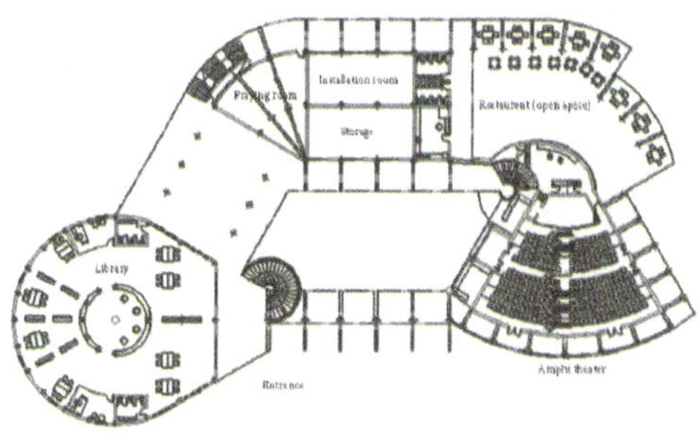

Ground floor plan

I used some ideas of light space of first collage, its grid ceiling and brightness passed through it, in my exhibition situated in the gathering space in order to harmony with the collage approach because the collage approach was lightness, coldness and well integrated with nature.

The significant subject in the first and second collage which I dealt with them, was a hidden space. It was felt easily but had no objective definition. It caused me to think about the hidden space limitless and apply my imagination force to materialize it. Finally, Amphitheater was hidden space.

In third collage, there could be seen reflection of a semi-open space along with a closed building behind it in the ground. I used the reflection of present mass in my middle space. So, classes were defined along the hanging porch. This part of design not only completed the rotating movement of space but also caused to show the massive space behind it more lightweight.

In order to close the spaces ring and according to the collages based on a lightweight and semi open space, the hanging porch around the central yard was closed and caused to connect separated parts of complex. Furthermore, the official and management spaces were designed second floor of it.

Finished the whole volume, plans and architectural documents based upon the existing architectural standards were completed. I analyzed different building systems and finally chose the cable system along with space frame structure. Details of small beam, cables joint and system of setting up hanging the porch on the cables were examined and designed by computer software to make the project as real as possible. At the final stage, model of cultural center showed better image of design. Reconsidering color, material and texture of collages, I sketched the resulted space to define the cultural center structure as well.

Cultural center design was prepared. I used collage as source of inspiration from start till end. Once again, my imagination and mind images came true.

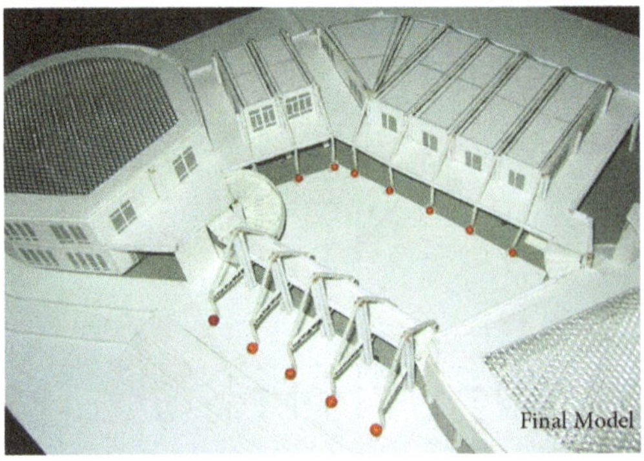

Site Plan

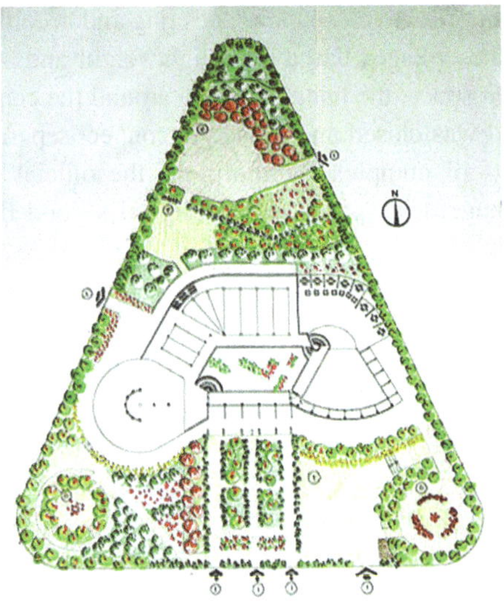

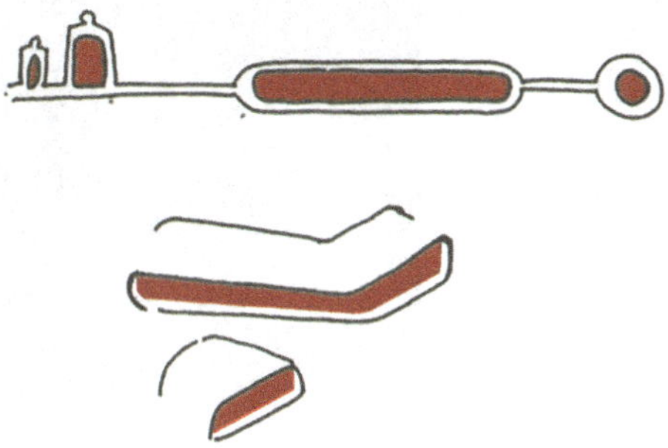

Up :
Bird View
Conceptual diagram

3.21 Moving Surfaces

Chizar cultural center
Student Hatef Ghahrodi
Masters Dr Ali Asghar Adibi
Dr Ahmad Ali Farzin

How is it possible to organize urban activity among work, business, leisure time as well to enjoy significant potential of private life and human scale with the nature? Architectural works depends on variety and complex factors of linguistics, global values, etc. In addition, it is possible to represent these works as white color by emphasizing on light using material characteristic and purity. Light, has special implication and usage, acts as decriminalization factor in the world, actually, there is correlation between it and nature.

Iranian architecture and traditional and Islamic architecture define reflection of a sustainable thought of a prototype that correlates past, present and future when creation activities have been vanished into some parts of vast horizons. Although, it is a controversial subject, it is considered one of necessary principle in Iranian-cultural design.

Earth is other subject. It is important to learn to live successfully and more sensitive related to earth, as a source of life. We need to change our behavior with the planet that creature are living. Instead of trying to dominate it, it is better to think about, respect and work on it.

Although dignity and reputation of our professional architecture has been destroyed by merchandise and bureaucracy, we are still able to improve the correlation between human being and earth fundamentally.

After short introduction, I will deal with some aspect of designing. Experience of architectural designer show that it is possible to reduce the energy level extremely by using the respiration layers of building.

Aerodynamics planes protect the building from the heavy winds, possible indirect movements and dangerous upward forces.

Articulated ceilings is open hydraulically This subject is considered a kind of dramatic aspects of design and make possible some parts of gallery multi-function.

I realized a fascinating color composition including white, black, red, jade green and during designing which acted on design creativity effectively.

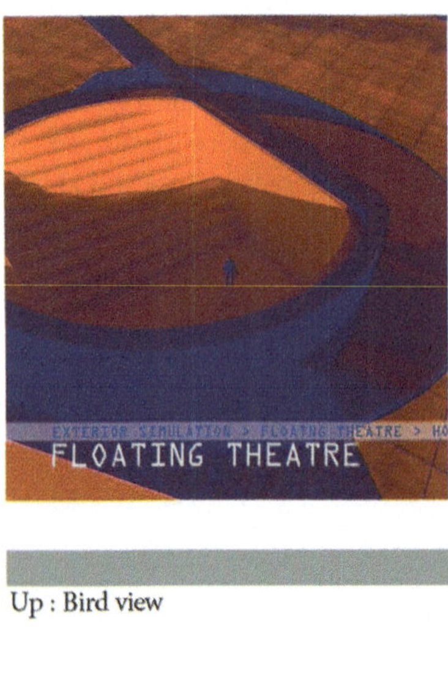

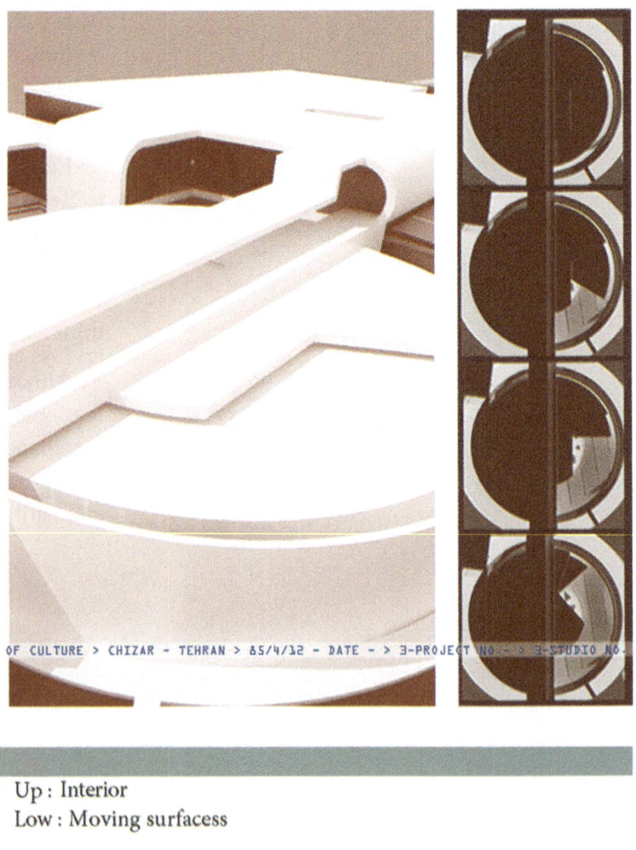

Up : Bird view

Up : Interior
Low : Moving surfacess

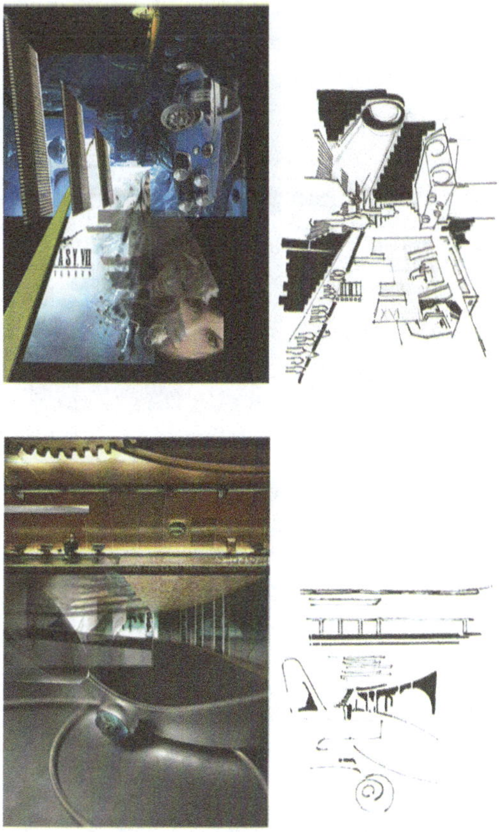

Up : First collage
Low : Second collage

The first collage

Design steps started with a stunning background. A blue background that showed unification and endless space. Correlation between earth and sky is established by a connection bridge and this sample is considered as evidence for it. A color balanced stair, heavy planes but weightless like feather, induce the feeling of ascension toward unknown point. The mentioned point settle in far away. Moving forward and fearless to concentration point. The concentration point of lights. Green bridge at the right side of scene emphasizes on subject more. The constant blue current, hanging projectors in path, variety architectural shells, downfall grids of building facades and vertical fraction define entrance and invitation.

The second collage

The most difficult case in collage is knowledge of each artist about nature of each piece but this causes him/her to do not know what the roles of each is. This subject may seem hopeless factor at first glimpse, but by watching the pieces in distance, artists can imagine whatever they want to know. At the beginning, my first imagination plan is some light radiations. This method of light dispersing on a plane is one of wonderful technique is obtained using a suitable light filter. Light halo at the left side of image is certainly sunlight penetrated through elements and set up on decoration fountain, it can be seen in second plan.

I cannot accept copper decoration on ceiling, instead, in my view the wooden agree with my design identity. Brass and copper materials show spaces more massive and heavier.

The green color of Plexiglas was not my interest, just for composition. Plexiglas induce feeling of lightweight and vastness. In addition, it establish more powerful spacial connections regardless of endless field of view. The up horizontal piece show a stunning sectional imagination and induce feeling of two floors architecture.

Therefore, horizontal movement at the top, rotating and deep movement at middle, and fountain in second floor are advantages point.

The third collage
Billboard and red, yellow and orange colors arose happiness and freshness. If there were not horizontal astonishing panels at the top, space seems like a nursery school. Nevertheless, cultural center and nursery school are somehow same as each other including colors, childish feeling of people when they enter a cultural center, cooperation sensation in performances, discussion and different cultural and educational activities for instance difference in colors and fabric.

- Orange decoration (ceiling fabric)
- Skylight and its continuous panels that emphasize on horizontal direction. Vertical elements such as low-rise yellow stucco semi walls are in accordance with horizontal path of top level
- Right side counter for instance reception or pre-space
- Distance between two neighboring pieces, an opening from inside to outside toward us based on billboard attraction and its direction continues at the end of picture and so imagination of staircase to upper platform. Generally, a first and second order lobby, a high division space.

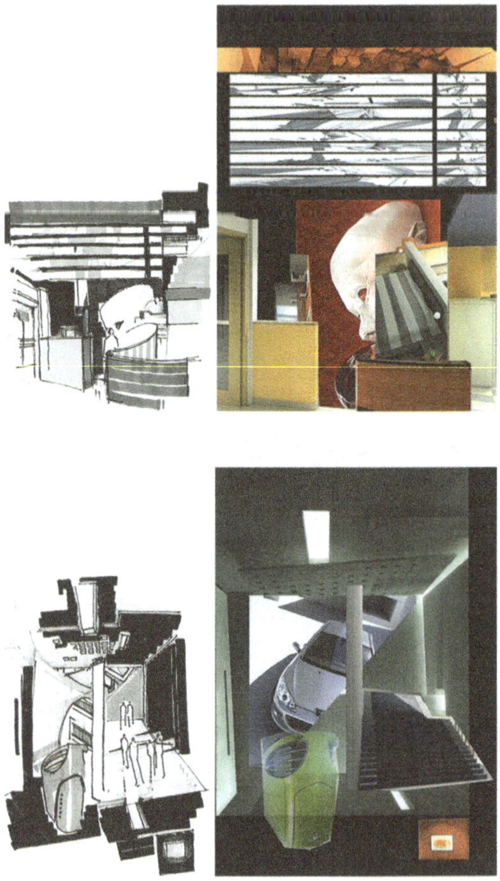

Up : Third collage
Low : Forth collage

The forth collage
I always think why light on floor that penetrated into space from an aperture is so beautiful; its value rise up in visual art extremely. It is result of hard working of sun and finally a great art! The collage started with slight color. The upper skylight induce two feelings: (1) Opening, (2) Building slab or hydraulic shell.

I observed the incomplete pieces sliding behind each other. These layers induce feeling of movement and sliding of these slabs.

Floor, walls and ceiling were integrated. On the other hand, they are at the same level. All are the same color, fabric and even lightness (rectangular light). This characteristic of sequence continue the whole design.

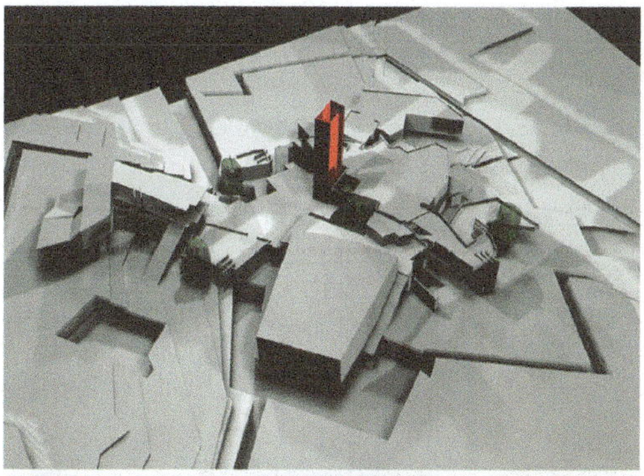

3.22 Illogical Sincerity

Chizar cultural center
Student Mahan Javid
Masters Dr Ali Asghar Adibi
Dr Ahmad Ali Farzin

At the beginning, I went to Arasbaran cultural center to understand the my design space as well and sketched some places that seemed wonderful. Then, I started to research and analyze the advantage and disadvantage points of cultural center and continued it thoughtfully.

I began to assemble some collages based on last semester (Thoughts sprang to my mind and caused me to choose special images. It seemed I heard slow shouting from inside myself).

First collage showed cozy and friendly space that its density, walls, ceiling and floor were the same size. Its characteristics would be explain more.

There were photos defined a high density wall covered by picture. In addition, cafeteria could be seen. Second collage was an unfriendly space which a red colored line overwhelmed its unfriendliness. There was an exhibition that its door and zigzag form walls set up near each other.

In third collage, lobby and educational com plex induce feeling of serenity and was colored attractive and stunning. In addition, I observed part of hanging and open exhibition, the zigzag form one, in this collage!

Last collage was ragged and unfriendly which I obviously saw the official part in it. A formal and authorized space with decorative walls. General space of collage form in my thought. Whereas I cut rectangular papers, I feel that all spaces are rectangular form. I began to sketch the collages. I had four or five sketch each of collage because I needed all side of space. Some collages even seemed different several days later. In next step, I started to make a model of spaces. The subject arose my wonder was a small model needed six picture because I thought they were pictures of a model from different sides.

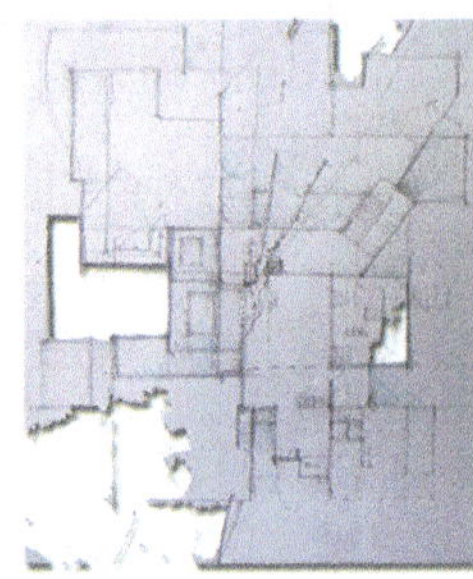

Finally, main part of design was formed. Next, I started to draw a plan based on model. Two lines break the regular spaces in two. I decided to put aside the regular approach, without trouble in sense of space, (I felt that this method is based upon collage more), so according to this the general plan was changed.

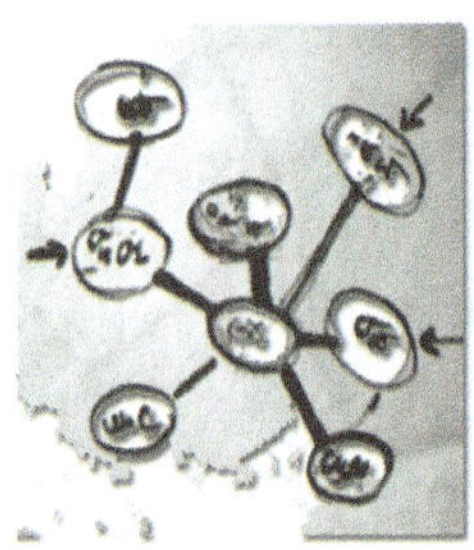

The design idea was based on two near zigzag lines, without natural light, the width of one could pass. There were pictures inside its walls. One find her/himself alone there and therefore has to see and think about them. A space at the end of path, is applied to represent pictures on its walls, ceiling and floor. It means passing by one dimensional space, you get to three dimensional space.

There is a high rise exhibition without stair, the only access way is elevator. Coming back from exhibition, a wall beside the glass elevator is full of pictures. So people can see new ones. Following the first exhibition, a corridor is defined as cafe exhibition. Thus, after ordering for drinking, person start to observe pictures and tools.

We began to site analyzing at last weeks. Therefore, a group of our classmates went to the site in order to observe it and study problems such as tree position, present building surrounding to access way, views, etc. Next step, I gathered information based on library books. Then I began to find solution for spatial and interior correlations. Knowledgeably, designing the space was done in best manner. First plan, for instance its spaces and windows position and height of different spaces, was changed according to gathered information.

Next step, site plan was designed. I extended all lines of plan, planes were acquired from the cross and parallel lines. Each of plane defined a position such as fountain, pool, park, restaurant, etc.

Finally, I made a model and realized the design problems during it, for instance to bring down the amphitheater in the ground, add several floors to exhibition, reduction of unsuitable fraction and locate the building on proper site plan. Following step, I organized the plan equipments and improve the slight problem. At the end the design steps were finished.

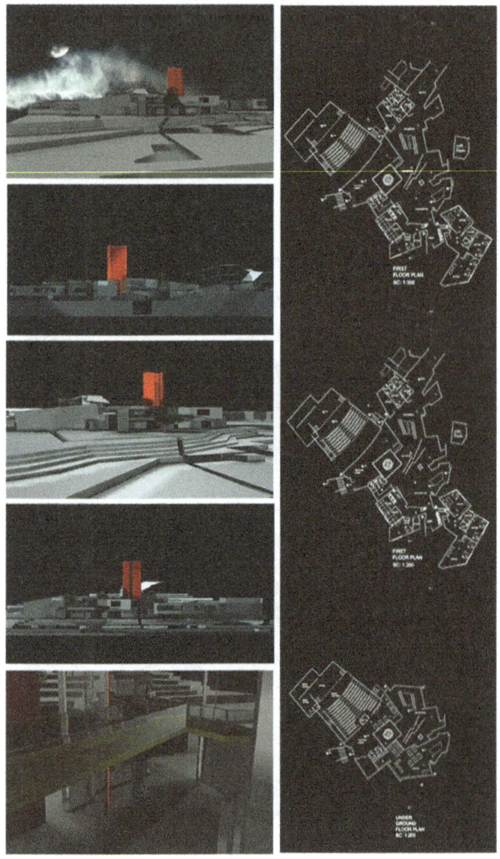

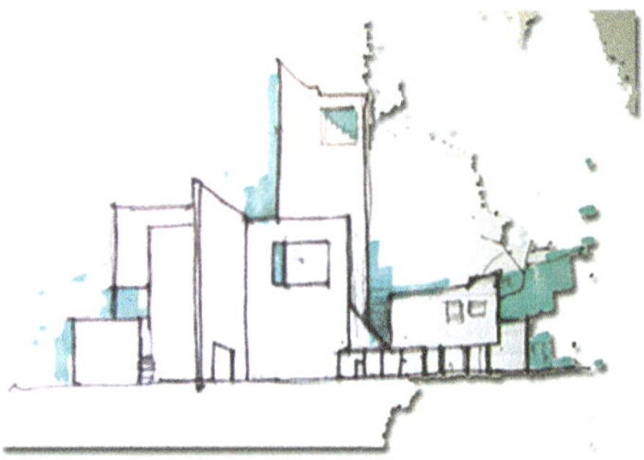

3.23 The Color of Grey Mind

Chizar cultural center
Student Negin Shakibi
Masters Dr Ali Asghar Adibi
Dr Ahmad Ali Farzin

I always look for a start point to do a work. This time is a cultural center! I should start now. But from where?! Plan? Section? Concept or ….

Vague images spring in my mind. Black, white, grey images. It seems I do not dare to color my mind spaces. I got confused among thousands of images spring in my mind. I should find a solution. Maybe the solution is collage. It may be a good start. A method that some reject it without trying it! But I tried it once and found the solution. Maybe it is start of wonderful end. Finally, I decided to experience collage again. Using the piece of papers form my imagination and color it. The unvalued pieces that gather together to shape my ideal space.

I flipped through the magazines one by one and cut suitable photos, rectangular form. I concluded to cut all of them in rectangular form because type of cut effects the space difference. Red, grey, orange, cream, white and black pieces. It was meant to compose the spiritless and cold colors and warm and fresh color like red and orange. I pasted the photos on a sheet and looked it in distance and replaced them with other photos to get a proper composition. The result was wonderful. Two different composition collages; But same feeling.

Colors of first collage were grey, black, white, orange and the second one were cream, red, white, black. Looking the collages vaguely, photos resulted in detail and accurate. Interior space of cultural center could be seen in collage. It was time to sketch these mental spaces.

The black and white sketches could not express the space same as collage. Making a model of interior space is certainly the correct answer. A model based on collage color and sketches.

The first collage, the grey and orange one, define a gallery that heightened 0.5 m upper ground. There is a black partition wall in the middle of gallery to locate drawings. Reflected green light of trees behind long vertical and opaque windows of corridor-formed space, at the right side of mentioned gallery, exhilarate the space. A column between each two windows present rhythm and movement in corridor. Natural light penetrate into space in this way and lighten one side of collage. This corridor would be entrance path to gallery.

The colors of left side of collage, grey and black, tend to warmer colors and space seems pale. Artificial lightening is used in this part and its lights glint and reflection of brightness on the floor and walls are obvious. There also was defined a slope path to gallery.

The second collage is full of warm color and its red colored parts wonder your eyes. Atmosphere of Space at the right of second collage, the cream colored one, is same as the left side of first one. Maybe these two parts are correlation of two collage. There is a glass space with natural light in the middle. This part is defined as entrance space, located lower than other. High red space, left side of collage, is narrow walls and heightens one meter upper than entrance. There is a partition above the mentioned space, that a string of decorative dark color spiral form is added to this partition. The connection path between this space and ground floor is stair. The created space define a gallery bigger than previous one.

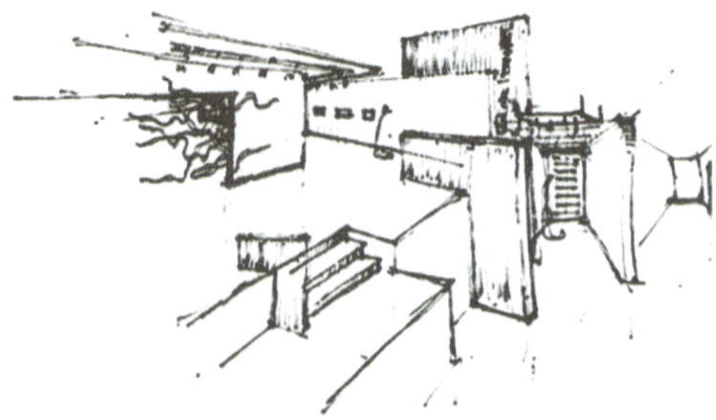

The light, radiated to a white wall at the end this gallery, is so attractive. This shows that space do not end to this collage and will be continue. Light of all these spaces generated artificially and reflection of everything and walls is obvious on shiny flooring. The spaces are create in detail in a way that all can walk on galleries and watch the pictures. Part of designing were finished when two galleries were connected by entrance. General geometry of designing was completed. Then I improved the geometry. The first space, lobby, was created based on the mentioned geometry. I continued designing by drawing a section of created space and kept on this section based on three dimensional view. To vitalize my thought imaginations, I tried to make a volumetric model. After model making, most of space were placed as well. General volume was created by two folded shells.

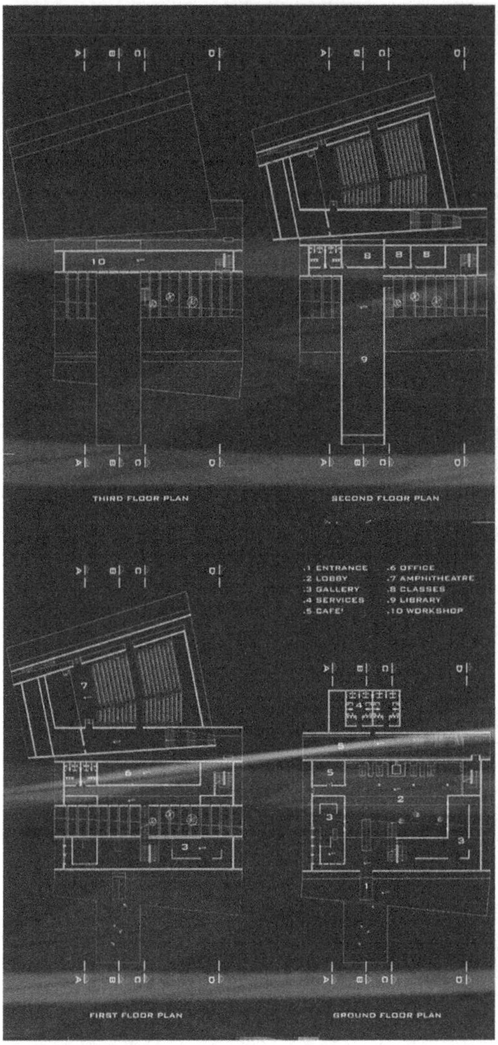

Galleries, entrance, lobby, official part and classes were organized in space made by first shell. This shell goes up in entrance, hangs up and again comes down to earth. I designed an open coffee shop and restaurant. The second shell created an amphitheater above ground level. Restaurant floor plane goes up, folds and forms a wall of amphitheater. Western and eastern walls of cultural center are covered by dark metal planes. There are grooved in the folded shell to have north and south light.

Lobby, gallery, classes and offices are supply by this light. I defined a cuboid library that its mass penetrate into first shell, pass it and finally come out from southern part and keep oblique and narrow columns hanging. The library shadow define entrance as well. There is a lawn space in the middle of shell, at the top of ceiling that huge and tall trees, in the lobby, get through it and come out of it. In addition, stair, started form ground floor, pass by this part of ceiling and end to library volume.

Spaces are completed now; colored spaces, places where all can walk on. The result is not only plans of building, but also lively cultural center. Spaces were created carefully using just vague pieces of images were gathered on a paper! The collage which give live to my grey thought.

WEST ELEVATION
SECTION A-A
SECTION B-B
SECTION C-C
SECTION D-D

3.24 My Scheme

Chizar cultural center	
Student	Nina Godasiaie
Masters	Dr Ali Asghar Adibi
	Dr Ahmad Ali Farzin

Third architectural design was cultural center. I did all steps based on collage same as last semester. Experiencing of collage attracted me and also the consequence of my scheme resulted from unconscious wondered me. I really interested in the pictures of collage without logical reason. Therefore, based on the experienced collage, I applied it again eagerly. I started to cut the images and pasted them in a way attracted me; the scheme which is for me; in other words, it is me. The images were colorful and I chose some just for its light characteristics. I began to prepare the collage. I had sense of belonging related to the four collage prepared sequentially and also enjoyed them. I found images in collages which resembled to the images seen in my childhood. I sketched the images of my collage. The sketched spaces looked similar to collage spaces.

Up: The first and second collage

Low: Third collage

The fascinating point is, the images resulted from collage were obvious, complete and contemplative. I gained a new space, using the resulted images from collage, which resembled to a restaurant and other space was a corridor.

I chose the best sketch and based on its color I made a interior model that represented the collage atmosphere. Then I improved it and maintained the best one, in the next step, upgraded other parts.

In my view, this kind of motion is admirable and contemplative. The movement which started from inside and get out. In fact, there is a continuous movement from inside to outside and vice versa. It is a reciprocating motion, which started from inside and is repetitive. I feel calmness, when designing improve. It is my designing and I sense it. Because the designing method, There is obligatory framework. Collage never induce constraint.

The important point in this method is, one's designing is particular based on his/her idea. All capacities are considered in collage, therefore it is a complete and perfect method.

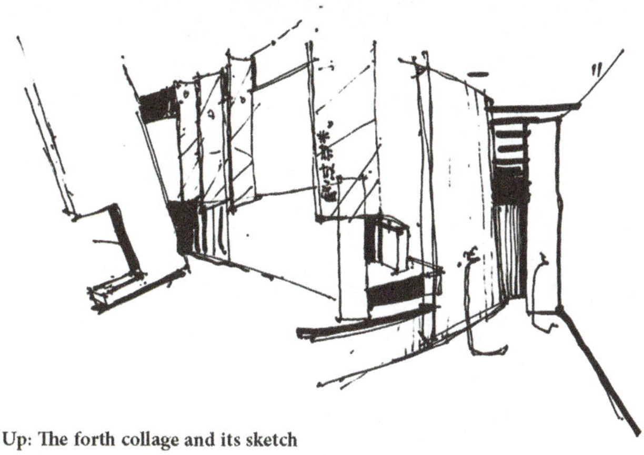

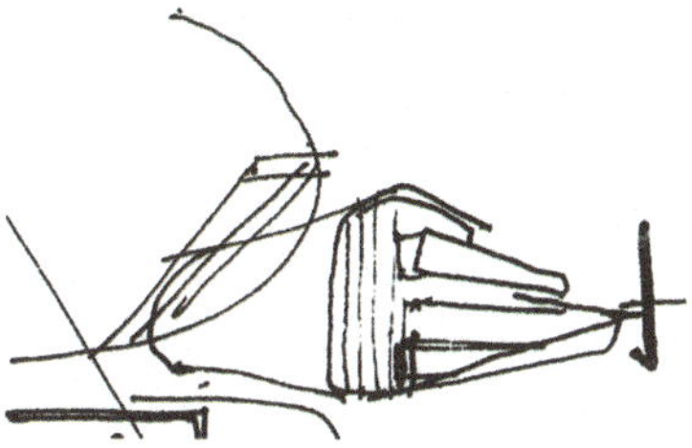

Up: The forth collage and its sketch
Left: sketch
Low: Model of collage

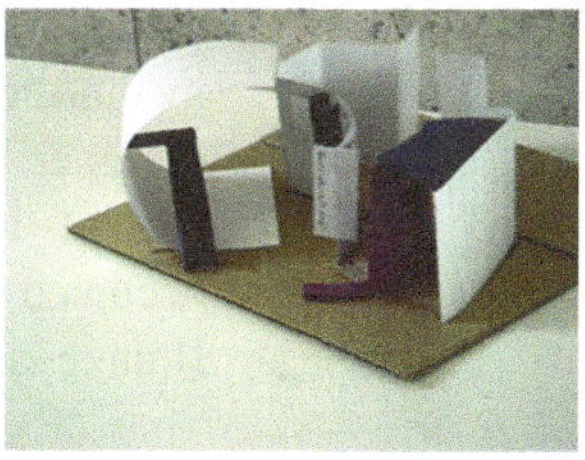

Up: The model of collage

Down up and right: sketch

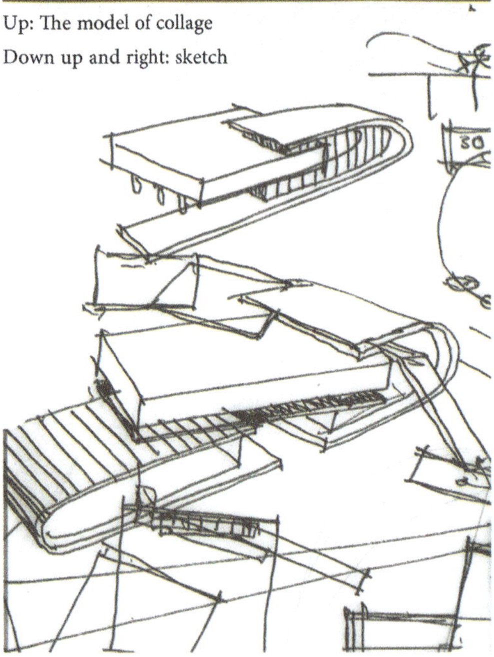

Collage represent a designing which I improve and sense it. It means that I walk into my spaces and imagine them using collage. I think about people attend in these space, how they feel visually. How high is ceiling and windows and what can I induce them using these space?

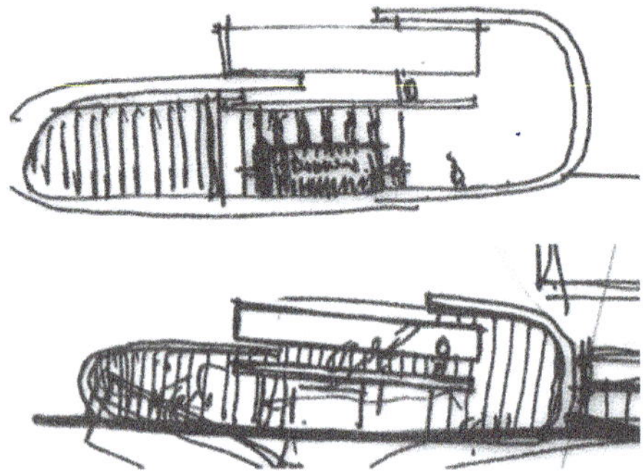

The glory of spaces is obvious; The open and high rise spaces. I started designing from inside and thought about outside. I kept my thought continuously. In my view, the result was attractive, it was a reference point which I pursued it, Because I saw the outer form in collage, a rotated and continuous shell. Its material, concrete, was even obvious in collage.

The collage talked to me. I found out about color, light, height, interior space, material and outer form using collage. It is not only the best method to gain a final design, but also help the designer to recognize his/her thought background. I have answered some questions which occupied my mind for a long time. I confront by strange and sudden subject but known during collage. The resulted images of collage reflect our mentality more than our expectation. If student apply collage, their work improve easily and guide them to gain the best result.

A motion from inside to outside and space understanding that achieving the best designing is the only result.

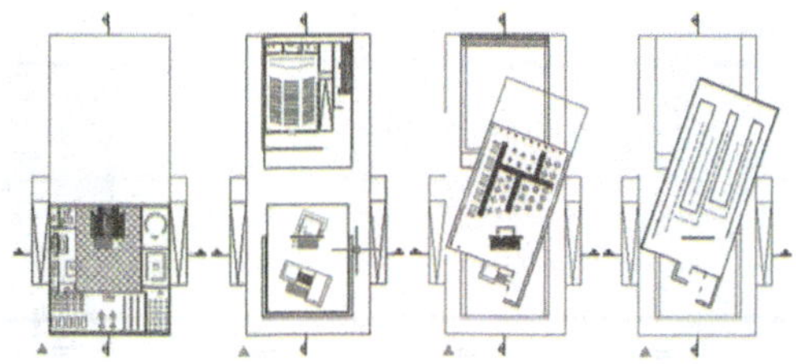

Plans of project

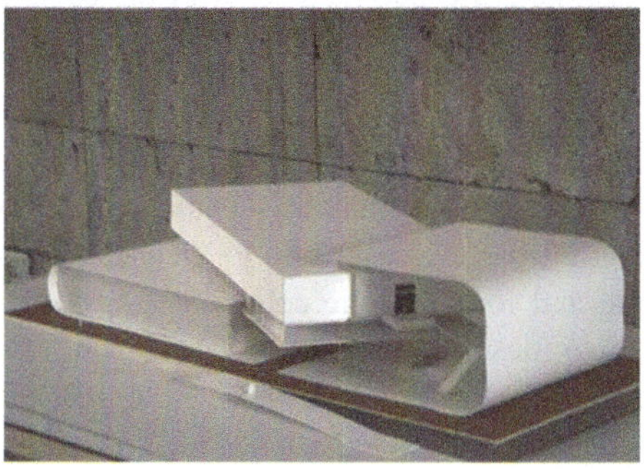

Collage understanding, as a important method, is space and its user understanding.

I ended up the designing and gained a wonderful result.

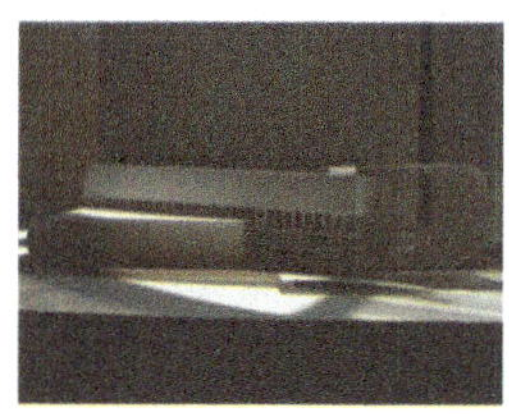

3.25 The Second Experience

Chizar cultural center	
Student	Ali Soroush
Masters	Dr Ali Asghar Adibi
	Dr Ahmad Ali Farzin

Designing condition was strange at the beginning of semester, because a residential building had been designed using collage last term, but this time, I started a new subject with the same teacher and method; a cultural center.

When we began collage last semester, teachers learned us to have building design based on your feeling and mentality, gather whole your interested pieces, according to their color, light, fabric and form, in one collage sheet. Is it possible to design a cultural center using those interested collage of residential building?

The main collage of cultural center

I started to prepare a collage after site analyzing and studying the teacher's desires. The only different between this and last term collages was: I chose the suitable pieces of cultural center and did not use all arbitrary ones.

The main collage of cultural center

I started to prepare a collage after site analyzing and studying the teacher's desires. The only different between this and last term collages was: I chose the suitable pieces of cultural center and did not use all arbitrary ones.

The model of collage

I paid attention during gathering the pieces. Unlike last semester, cutting lots of images from magazines, I chose the pieces carefully and thoughtfully this time.

The resulted collage, after preparing some colleges, showed that creativity in collage is important factor in interior designing.

The model of collage

My collage, with its colors and forms, defines a feeling which is not possible without it. When Dr Adibi (Group supervisor) provided the sketches of my classmates and Tabriz university students of my collage. I found out that they are designs based on my initial principle of idea, it made me happy.

I sketched the collages; drew a plan; and made a model.

This was undoubtedly the best experience for me in last semester.

Top: Sketch of main collage
Right: Model if collage
Below: Site design

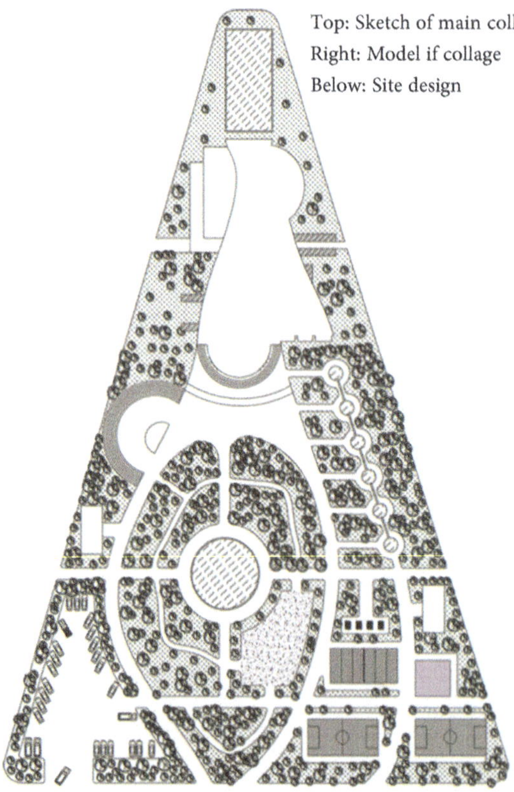

When I realized my sense of local cultural center was formed in final design, I got close to my architecture. Light and color of model and sketches caused me, same as adult, to walk on the my ideal cultural center and experienced the childhood and adulthood my best memories.

Top: Second collage
Left: Model of collage

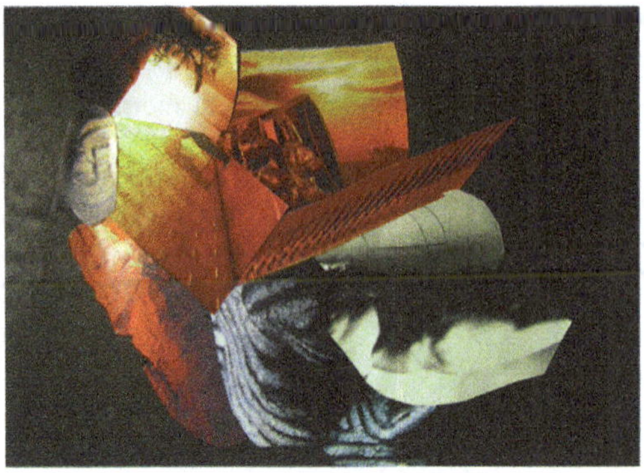

Top: Collage
Below: Model of collage

3.26 «Collage», Second Sequence

Chizar cultural center
Student Farhad Binazadeh
Masters Dr Ali Asghar Adibi
Dr Ahmad Ali Farzin

The purposes of reselection of collage, as a design method, in this semester were to know it better and completion of previous experience, although there was difference between this and last terms: the importance of outer form of new design.

Unlike previous term, first step, preparing collage, progressed rapidly, without confusion. Therefore, sketching and collage was purposeful. Design began by preparing two collage. I tried to make a space instead of sketching, so I made a model after collage immediately. In addition, colors was applied in model making. Like the previous term, the initial model defined the resulted sense of collage not the architectural space.

Fortunately/unfortunately, I came across a problem in designing which was a great personal experience. It must have helped me to know this method more.

The problem was, I lost the early collages.

As regards I had a model based on collages, my teacher guided me to restart and prepare new collages. I thought that making a collage depends on time, but when I ended up new collages, I found out these ones are the same sense of previous collages, however more developed.

I made a model based on new collages immediately but paid more attention to architectural space because the short time. I even tried to represent the outer volume in one of model.

Like previous term, I designed spaces such as gallery, amphitheater, etc. After that, the designing went on more accurate.

Applying the collage in plan was the subject helped me to design the plan and organize two dimensional lines. Finally, I realized the rotation angles of lines of plan, same as collage, are 18°, 34° and 90°. These lines were used as guidance in site design; actually, this geometry was not as a limitation factor. In addition to organizing the design, it was a guidance element which help me to get closer to the purpose, defining the sense of collage using the spaces. At the end, the result was satisfying.

Collage

The model of collage

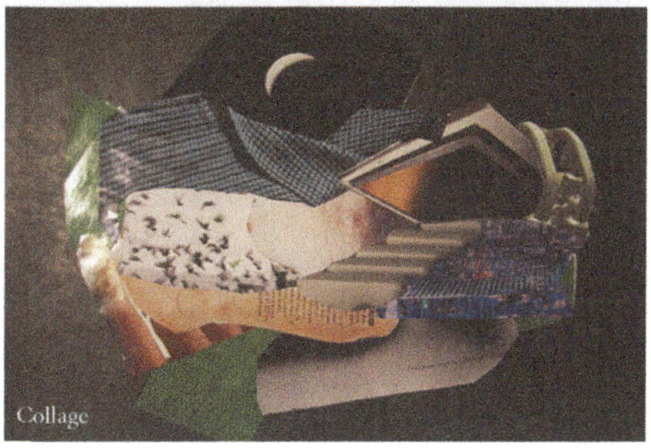
Collage

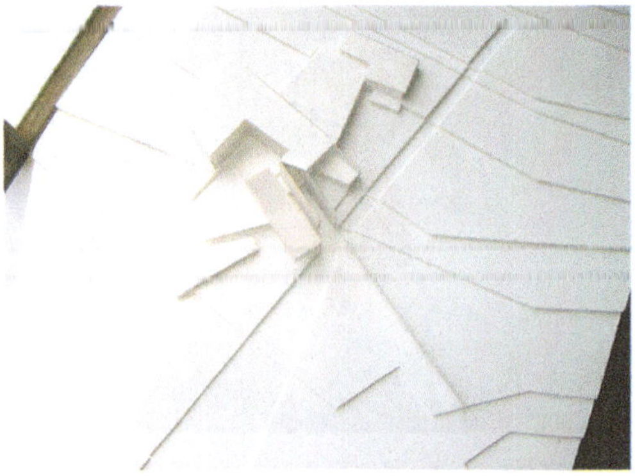

I experienced by the new design that collage is a starting point and guidance for architectural trend. Based on one's creativity, approach of people toward space is different. Even the results of two persons designs, based on their activity trends, are different for a collage.

My experience of collage was improving during two design classes. First term for getting familiar it and its usage, but creativity and personal approach is considered in the second one.

In my view, collage is significance method as a starting point. When our conscious minds are not able to choose the best method to begin, we can use our semiconscious minds and interior abilities. Collage is the most suitable method to start and finalize a design.

Top: Final model
Below: Model of collage

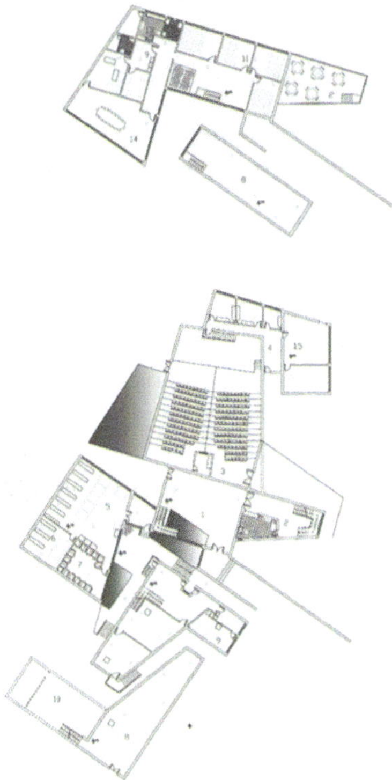

1-Lobby
2-Coffee Shop
3-Theater
4-Back Stage
5-Library
6-Refrence Book Room
7-Computer Center
8-Gallery
9-Office
10-Gallery of Statue
11-Classroom
12-W.C.
13-Service
14-Confrence Room
15-Kindergarden

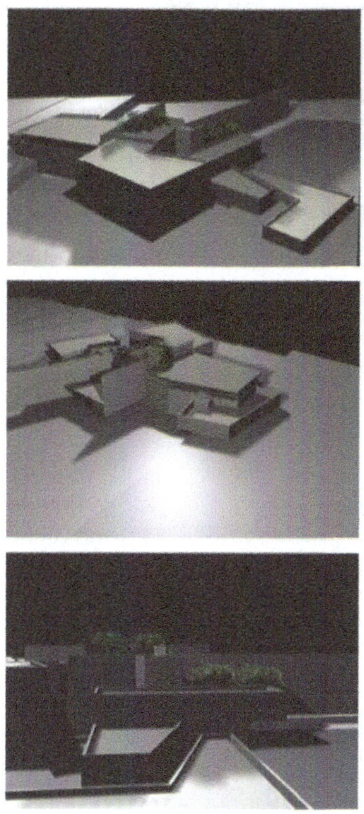

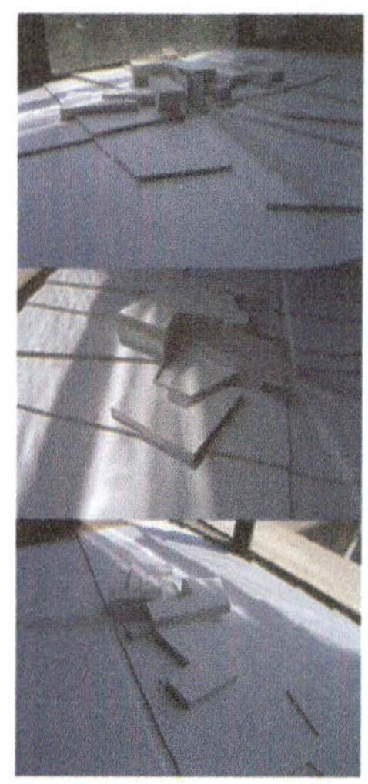

Top right: Computer base model
Top left: Final Plan
Below right: Final model

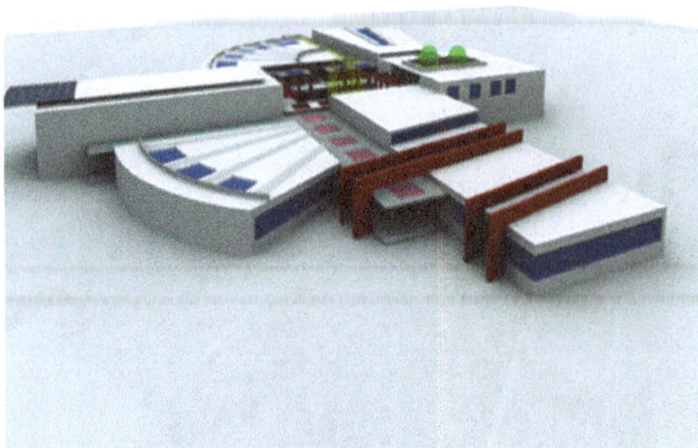

3.27 Large Scale «Collage»

Chizar cultural center
Student Laleh Reshad
Masters Dr Ali Asghar Adibi
Dr Ahmad Ali Farzin

Chizar cultural center was the subject of this term designing. The scale of mentioned design is larger than residential building one. Circular form, defined the entrance, was first element which used in collage. When I observed this form, I felt that a buffet or mall should be behind it.

Pasting several images on sheet, it formed sequentially and induced sense of depth. Dominant colors were tones of dark brown and black. In order to enliven the spaces, they needed light. I decided to define low depth yellow space behind the curvilinear volume and after that design a friendly, not a formal, space. A space that excite feeling of proper coldness, acceptable light and ideal silence, or stunning light which absorbs and invite inside. I was interested in gaining a semi-open green space, defined at center of spaces. In this step, I added spaces a image which induced a four side space, surrounded with rooms. A space which might define the mentioned green space.

Collage was divided in two parts including, yellow and green space with stunning light and the other is dark and brown space.

Also spaces were created behind the mentioned space. Intensity of light and long corridor absorbed me inside. I walked on spaces of collage. If I took straight path, I might arrive to blue space, induced a pool, or stairs to mezzanine. When I went to down stairs, one way guided me to vast hall with stunning light. There was a curvilinear space at the right side that seemed to define end of space.

Design process

At the beginning, I started to make a model. Colors were organized beside each other same as model and I defined spaces based on it. A stair-formed space and its long corridors caught my eye. Gradually spaces were defined by organizing the space beside each other. New spaces were formed at the center of spaces. Distances between these spaces and corridor and stairs should be defined. I decided to work on each spaces and found the function of them.

In this step, different spaces of cultural center were defined and attached each collage space a function. I work on spaces separately. I started with sketches of collage and tried to complete them by model and collage. In doing so, whole spaces were created in my mind and I just organized them. I reused the collage to realize the location of spaces. The resulted structure seemed horizontally more than expected. So I relocated and put them behind each other. At first, the square stair absorbed visitors. There were a curvilinear form at left side and depth space at the right side.

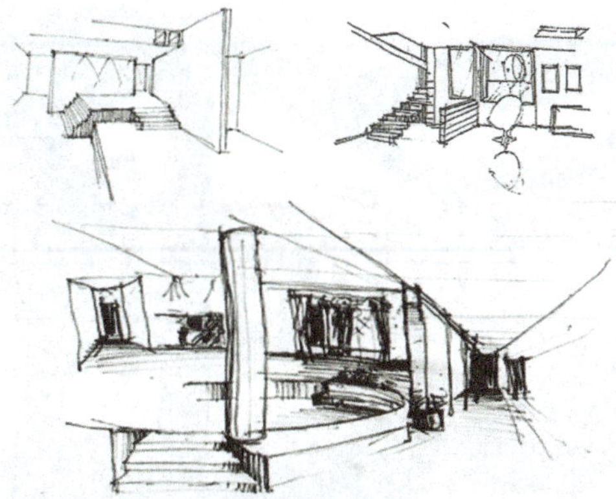

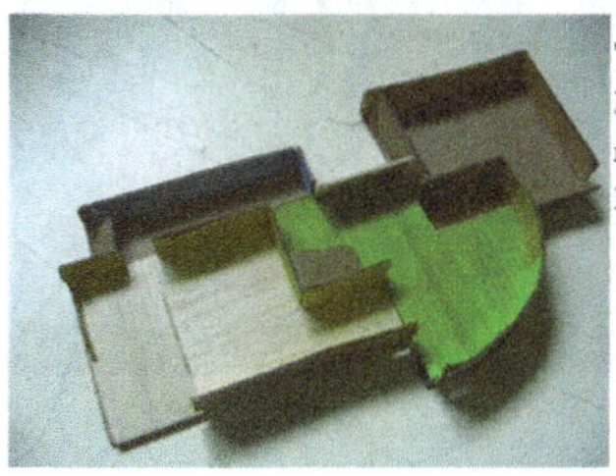

Tope: Sketch of collage
Middle: Computer base model
Below: Model of collage

Computer base model

Moving forward, a semi-open space defined as central point. Spaces such as gallery, classes and amphitheater were organized surround it.

The central part should seem more noticeable. So, I made a decision to design its ceiling differently and tried several types of glass ceiling. Finally I arrived at a conclusion which it was possible to define a movement way (corridors) in this space. A garden gradually was set up in the center of space that waterway flowed toward it. The idea of glass staircase came to my mind; a glass wall with stair and fountain surrounding it. To have unity among spaces and because the spaces were organized separately, I decided to apply ceiling in whole spaces. Each of them was defined by colored glass in ceiling.

Spaces description

Central space
Looking to the center of collage, it reminded me a glass or blue plane. This caused to apply a glass plane and fountain at the center. The ceiling of space, because it considered as prominent place, was different. There were sitting platform at the edge of space. Garden surrounded by gathering places.

Class
This part was a four side space with rooms surrounding it. This design was resulted by left side of collage.

At the beginning, I decided to set it up at the center, but located it beside based on its position on collage.

Site design
To strengthen the sense of main entrance and provide a gathering place, I planned to create a vast step form space in front of the structure. Moreover, the southeast axis were powered by defining squares and ribbon of water. I designed a place for gardener at the end of site. Space such as buffet, cozy place and playground for children were part of site. Also, there was an open amphitheater at the top part of site.

Gallery
Gallery was somehow glass made. A garden in the middle divided it in two. Green and cream yellow spaces gave sense of freshness and coolness. A skylight was defined at top of garden. The space form strengthened movement toward open space. This part was designed base on left side of collage.

Playground for children
This part was resulted from curvilinear form at the center of collage. A form which ascended step by step. Children could play, draw, sit or make a model at the center part or between steps.

Library
A light and deep space could be sensed at the center of upper part of project. This space caused me idea of a long cubic space next to entrance come to my mind. This space was for library, sub gallery and a drawing place. I was interested in defining a long windows for library.

Top: Sketch of collage
Below: Computer base modeling

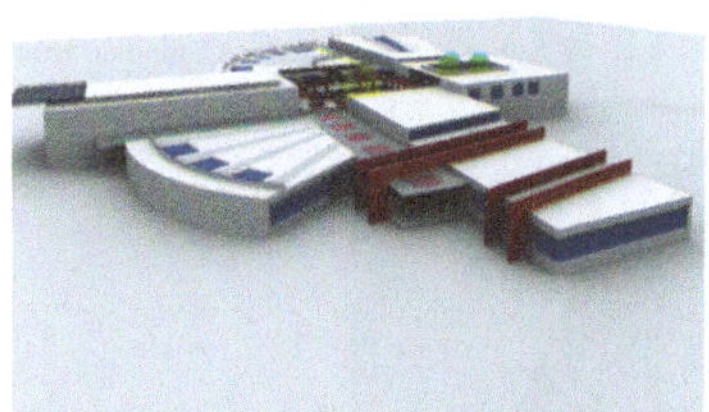

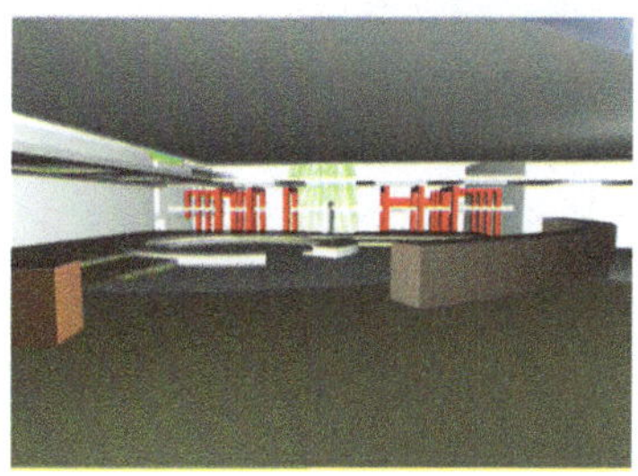

Top: Second collage
Middle: Computer base modeling
Below: Sketch of collage

Second collage

The second collage was based on curvilinear form, but it seemed as a medium and division space this time. I felt the entrance is in the right side of image. The light, penetrated into space, intensified this feeling. There was a formal space in the left side that seemed a conference hall. Door at the right side of stair guided us toward upper level. A long corridor with warm and red colors and prolonged windows directed us toward the end and exterior space.

1 Entrance
2 Restaurant
3 Library
4 Amphitheatre
5 Central space
6 Mall
7 Playground
8 Gallery
9 Minor gallery
10 Offices
11 Rest room
12 Installation room

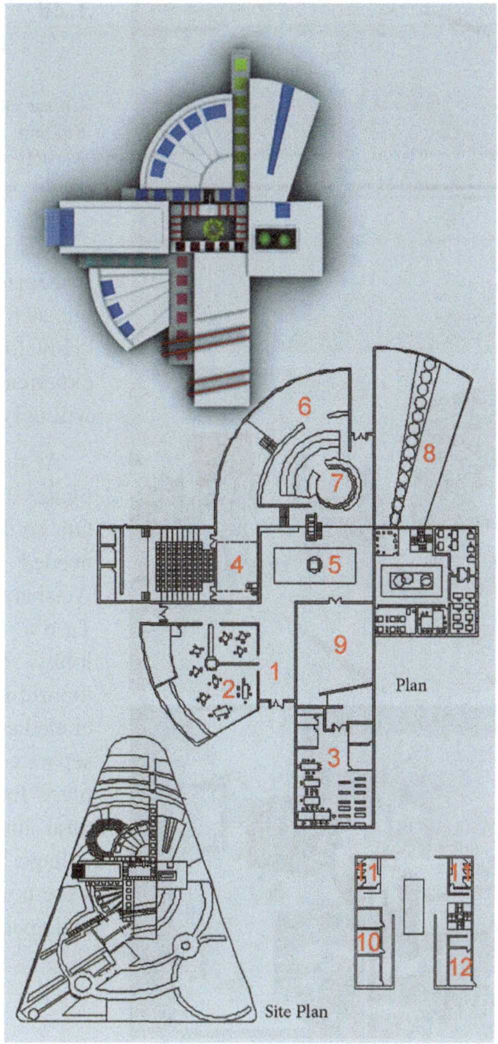

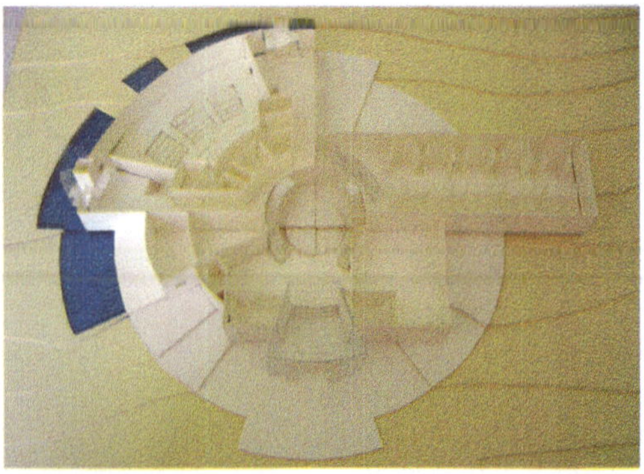

3.28 Reinvestigation in Labyrinth of Mind

Chizar cultural center

Student	Nilofar Zarnegar
Masters	Dr Ali Asghar Adibi
	Dr Ahmad Ali Farzin

… Restart

It seemed no vague this time. We knew the path, but at first "what to arrive" was not important. "How to arrive" was considered. The path which experienced before and would be examined again willingly.

At the beginning, It was important to investigate some information about cultural center. Our collages must be our cultural center but mind needed sparkle to start. So we decided to go Arasbaran cultural center one day before day of Taso'a. There were painting of Ashura in its lobby. Passion readers performed Tazia and mournful people gathered around them. I concluded cultural center is not considered as several separated spaces with its functions; it is gathering place for people have in common such as cultural, historical and religious. I began to made a collage based on these ideas and thought. They were not as residential building. Its scale, colors and space feeling represented a cultural center.

First collage
Personal calmness and privacy dominated this collage. There is red beam at the corner and their soft reflection can be seen in blue backgrounds. Forms were combined and colors were mixed together quietly. Horizontal and vertical border lines are visible and it is small scale.

Second collage
This collage is full of attractiveness and it seems each part tells something. Transparent frames at the right, obvious and curvilinear motion at the left and a hidden softness in straight lines at the center. At the beginning, I could see the exhibition in collage. A dynamic one point perspective along with a pure curvature at the front …

Third collage
There is motion in this collage in a manner that can be felt in each side. Colors combine and masses are organized beside each other. The image is full of curvilinear and reflective motions.

Forth collage
Similarly, this collage is full of dynamic movement. Combination of rigid and transparent volumes, curvilinear with straight line forms and finally a frame-form and different motion at the center.

Second collage

Third collage

Forth collage

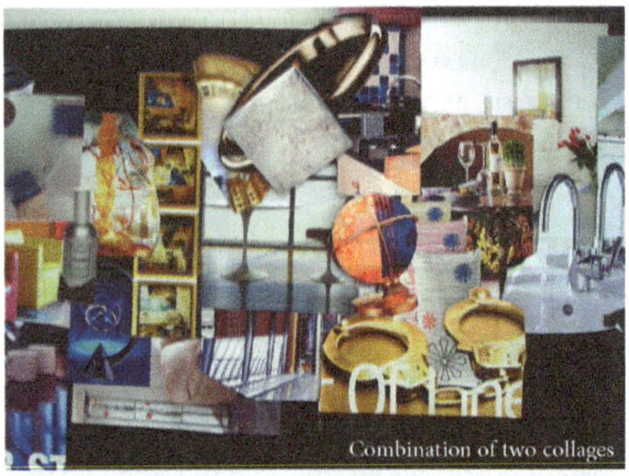

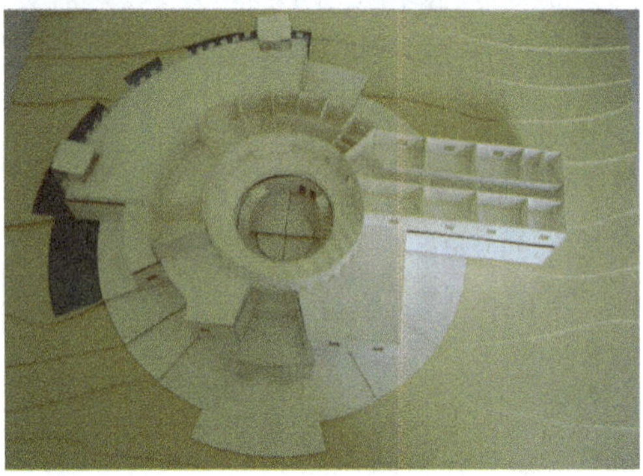

Combination of two collages, a fresh image

The two last collages were full of important subjects, but I cannot separate their spaces in my mind. It seemed invisible strings linked them. Their common characteristics caused me, after consulting with my teacher, mixed them. It was amazing, it seemed the result image was unite from the beginning. The unity between images leaded to formation of lobby and general geometry in my mind. An empty space, shiny and circular, at the center and a space with difference in its height under it and kindergarten and playground at the left and finally main entrance at the right …. This image formed the general lines.

… and the third designing

I observed the site location and analyzed it at this semester.

The general idea was formed in our minds. We applied them in paper, sketched the collages and made models. Providing the design necessity was following step. The design was done step by step and our teacher guided us.

The designing was over. A remarkable experience of designing, same as previous one.

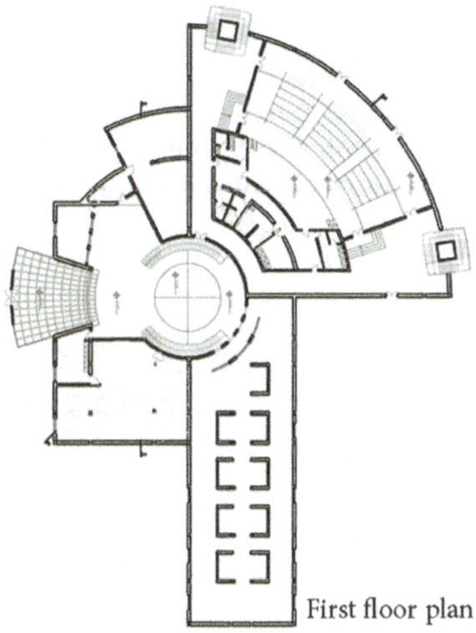

First floor plan

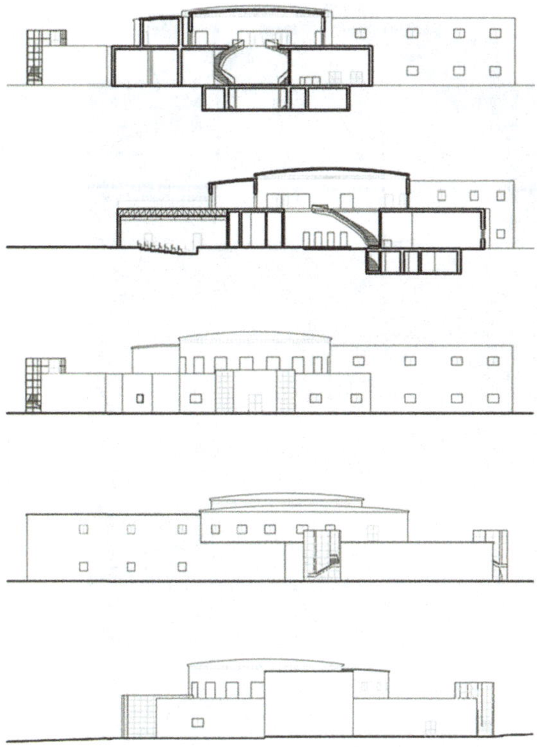

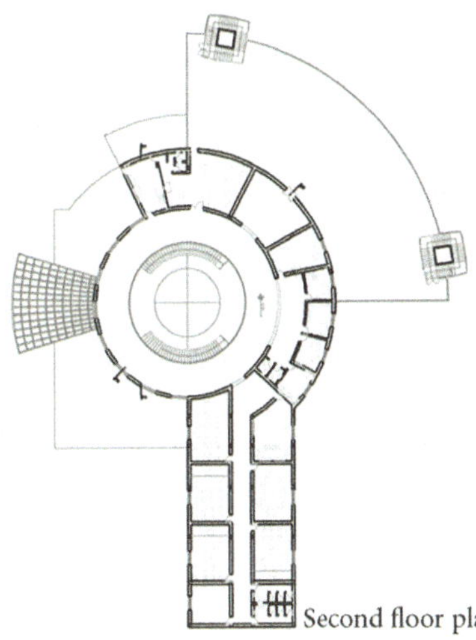

Second floor plan

3.29 Spatiography

Chizar cultural center

Student	Parisa Torabi
Masters	Dr Ali Asghar Adibi
	Dr Ahmad Ali Farzin

… A restart
Bahman (eleventh month of the Iranian calendar) of year of 1384 (in lunar year). Sixth semester. Third architectural design. The subject was designing of a local cultural center located at site in Chizar-Tehran. I was always waiting an opportunity to represent my skill in this kind of design.

What should I do? Should I apply past method so as to examine my ability? or should I use collage again and represent a design to experience a new method?

I chose the second one and started designing.

First step: endless research;

What is cultural center? What is culture? Should I design based on culture or its appearance according to nowadays and its inside based on original and traditional actions?

I had to research and preferred to study built structure. We (along with our teacher and my classmates) went to visit two local cultural centers; Shafaq and Arasbaran cultural center. I was so effective. I walked on these space unlike before. I was not same as theatre audience or music listener, I should study interior space program and provide the great situation. Visited the site, I had to analyze results. This step is important and time-consuming. But I did it greatly.

Second step: Being at real action;

It was time to begin practical part. I knew somehow what I should do. Hence, I gathered whatever were needed such as images. At the beginning, I looked them over and over and each time got new point. I gradually cut images. The ones which formed my designing. I promised to privilege collage in all spaces designing. In doing so, I had to prepare collages in manner that answer whole design necessities. I stayed loyal to this action and it was so valuable.

At first, the cut images seemed inharmonic and irregular. But after short time, I realized it is possible to categorize them based on color. Arranged them in group, I found that there is pleasant harmony between images according to their colors or lights or even space atmosphere. The project necessity was provided and I had to began designing.

First collage

The first collage should represent main entrance. I tended to do it step by step so as to not to get confused. Fresh images, red color ones which imposed cozy spaces, was used in this collage. Finished the collage, I looked it from distance. A cylindrical space could be seen in the middle of collage. It was entrance with a long windows. After it and passing a level difference, we can get into main space, a friendly and cozy one. Place to standing, sitting and getting familiar. There is a fountain at the center. Light beam behind it represented that something relation with cultural center was happening.

Squares could be seen at the bottom of collage which showed regularity of rotating geometry of entrance. Hence, program happened inside it, seemed important, such as visiting of exhibition galleries on the part of artists and art lovers.

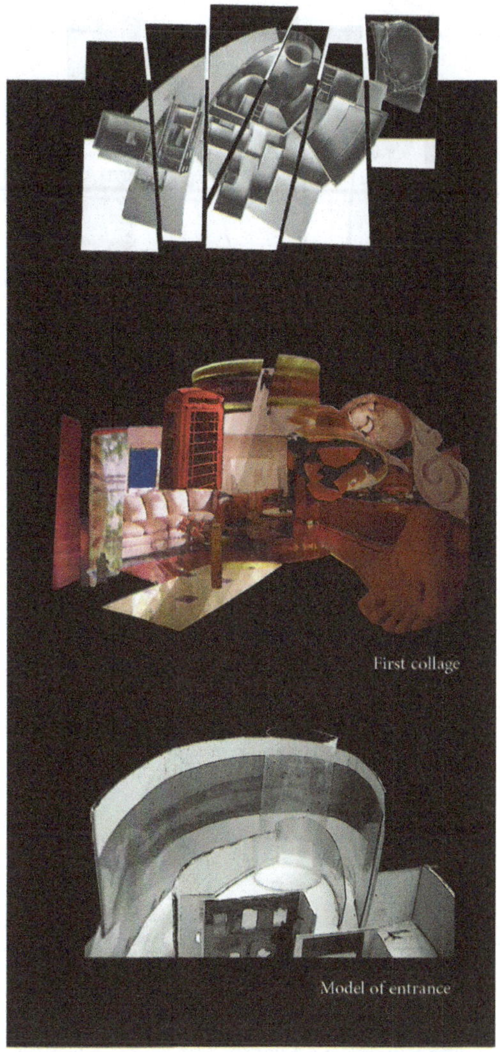

First collage

Model of entrance

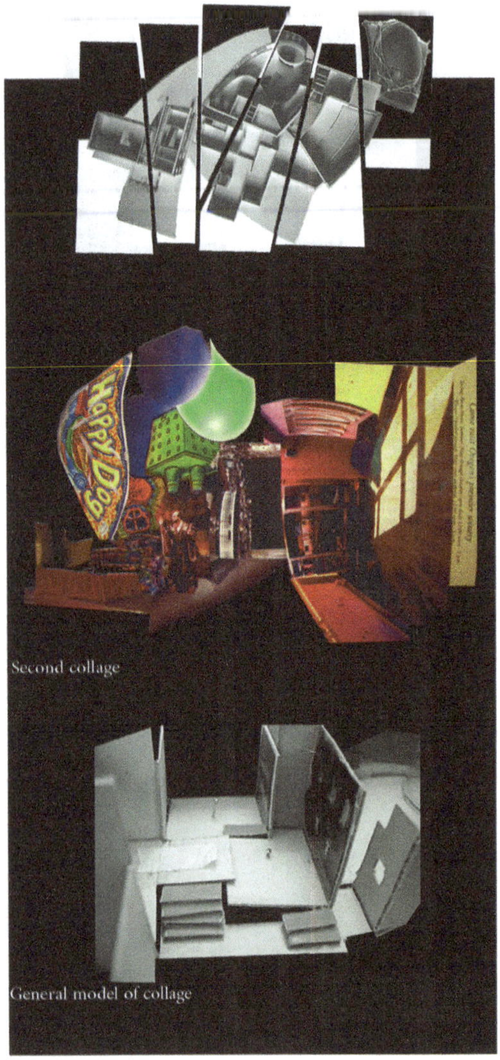

Second collage

General model of collage

At the right side, passing by a curtain wall, there was a space behind and next to entrance which is out of sight. It was suitable for spaces like restrooms or secondary paths for exit and entrance.

Image, at the right side of collage, ascended curvedly, guided us toward upper level. Hence, this space should be defined in two levels.

Second collage

Images were dark this time and narrated spaces without relation with exterior. I preferred to make a collage for amphitheater. I gathered the images and looked them from distance. I realized different space each time. I sometimes recalled long years ago but sometimes found myself in nowadays atmosphere.

None of them satisfied me but I achieved better result and combination after period of time. The result was neither modern nor monument. Instead, it was a space based on traditional root. Moreover, it imposed freshness and novelty. Its colors not only, beside its at most transparency and clarity, did not reduce the space importance but also it gave space special atmosphere. Interior and exterior of hall seemed different. In doing so, to get into the hall, we should pause and prepare ourselves, then enter it. A space was defined to pause so as to separate the visitors from outside and next get into it. Walls were transparent in order to people do not feel themselves in other world suddenly. Part of ceiling was transparent to penetrate light.

Generally, amphitheater was formed based on cylindrical entrance and the main curve. Therefore, it probably was defined beside main space.

Third collage

Different entrances was one of characteristics considered for my cultural center. I did not tend to define an entrance because visitors felt that there is one way to get into space. Hence, beside main entrance, more ones should be defined so as to not to concentrate on main entrance and emphasize on other spaces. Moreover, it caused to provide more access way.

So I prepared other entrance collage. The chosen images had to represent all matters because a lot might happen outside the cultural center. I tried to use images irrelevant to subject to find out the desirable space abstractly. The important point was presence of people. They had to define whatever happened outside. However I applied personage previous time accidently, but its usage was deliberately this time. To show people gathering, a lot was used.

Fresh experience in each step of design is privilege of collage method. Designer think about one part of design anytime and assume each part new matter. This characteristics cause the designing continues from part to total and conversely. Hence, it grow regularly. While other method improve without consideration of special characteristics of space or main space often is designed and other spaces are attached based on it. In doing so, all spaces are considered important according to main one. This means "annihilation of space values".

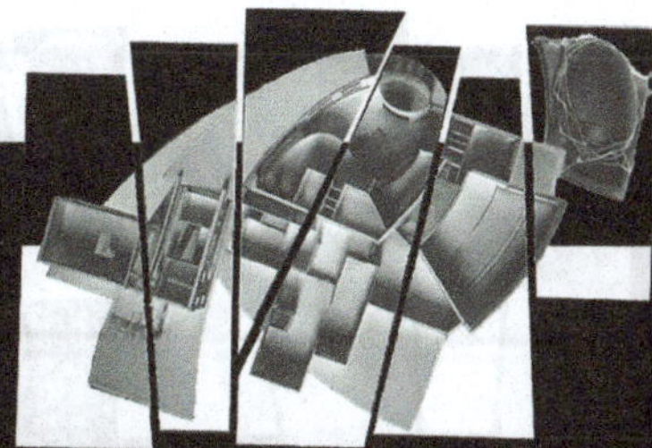

Third collage

Model of minor entrance

Forth collage

Model- Interior to exterir view

This process never happen in collage and each space has its value and protect it, as I achieved my ideal space in second experience. This entrance located behind amphitheater, so visitors could get in intermediate space then to the amphitheater and conversely. There was a gathering space, pavilions, next to entrance to represent handicrafts. A joyful and fresh space was provided. Galleries which located behind pavilions on a platform, surrounded by platform and main space (as it was expressed in first collage).

Fountains and element of landscape design could be seen around site which improved beauty.

Forth collage

It was necessary to think about forth and last collage. I wanted to prepare a collage irrelevant to subject. So I thought and reviewed results of previous collages. Most spaces were designed and main geometry was obvious. I almost realized relation between cultural center and landscape. Site and its design were last matters. The subject of forth collage: Designing of site landscape.

I applied abstract again. Images were chosen based on their color and light. I put them beside each other several times; different arrangement. Each seemed beautiful. Finally, I chose the one which was based on cultural center geometry more. Curved lines, intermediate between outer shell of structure and outside, represented curvilinear geometry of interior. Too bright light showed that surrounding of site was open and light penetrated interior space without barrier.

I used personages again, people who gathered circular and looked at one point. In fact, the geometry of circular open air amphitheater formed a complex which due to its distance with building, located at the top part of site and its connection with all complex was provided by curved path. The site designing was almost done and following step started.

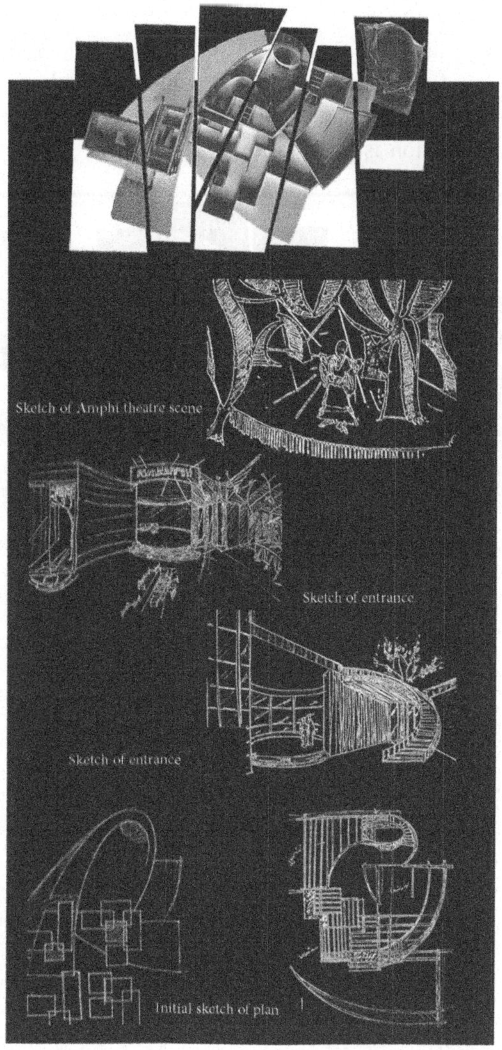

Third step: … main part

I had to gather previous results and form generality. In doing so, I sketched collages, whatever felt or observed in it.

This helped me to continue more seriously and realistic. Next step was vitalization of space which one part of designing was considered as starting point, then I made model of its space. However, the model was scaleless or proportionless but guided me to continue better. I imagined several spaces beside and made second model. It improved more realistic step by step. I had to continue scale based using metric handbook. Third model was made according to standards and scale. While the site slope was 5%, so I had to locate the model on it. Therefore, I prepared site and place the model. In this step, I could tell "the designing is running out". Detailed plans were prepared. Consulting with teacher over and over, I changed the plan and model. Finally, designing of Chizar cultural center was done.

Last words
We examined ourselves by designing of a complex. I experience new matters and this made me happy. Designing by collage. However, space assumption is better title.

I especially appreciate our teachers to allow me a chance to achieve one of the best during education. I hope to have chance in future to offer it to other student and they also use this valuable method.

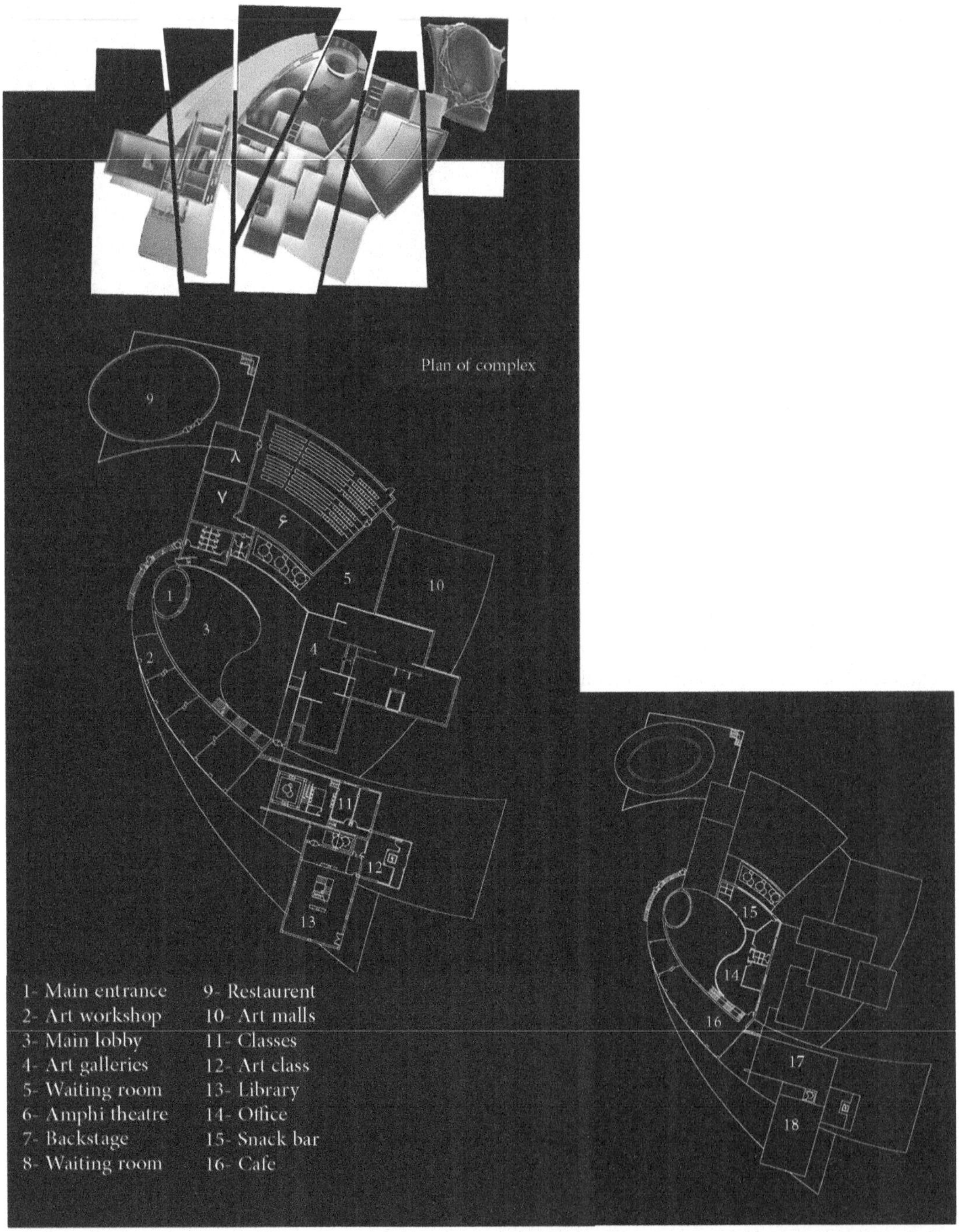

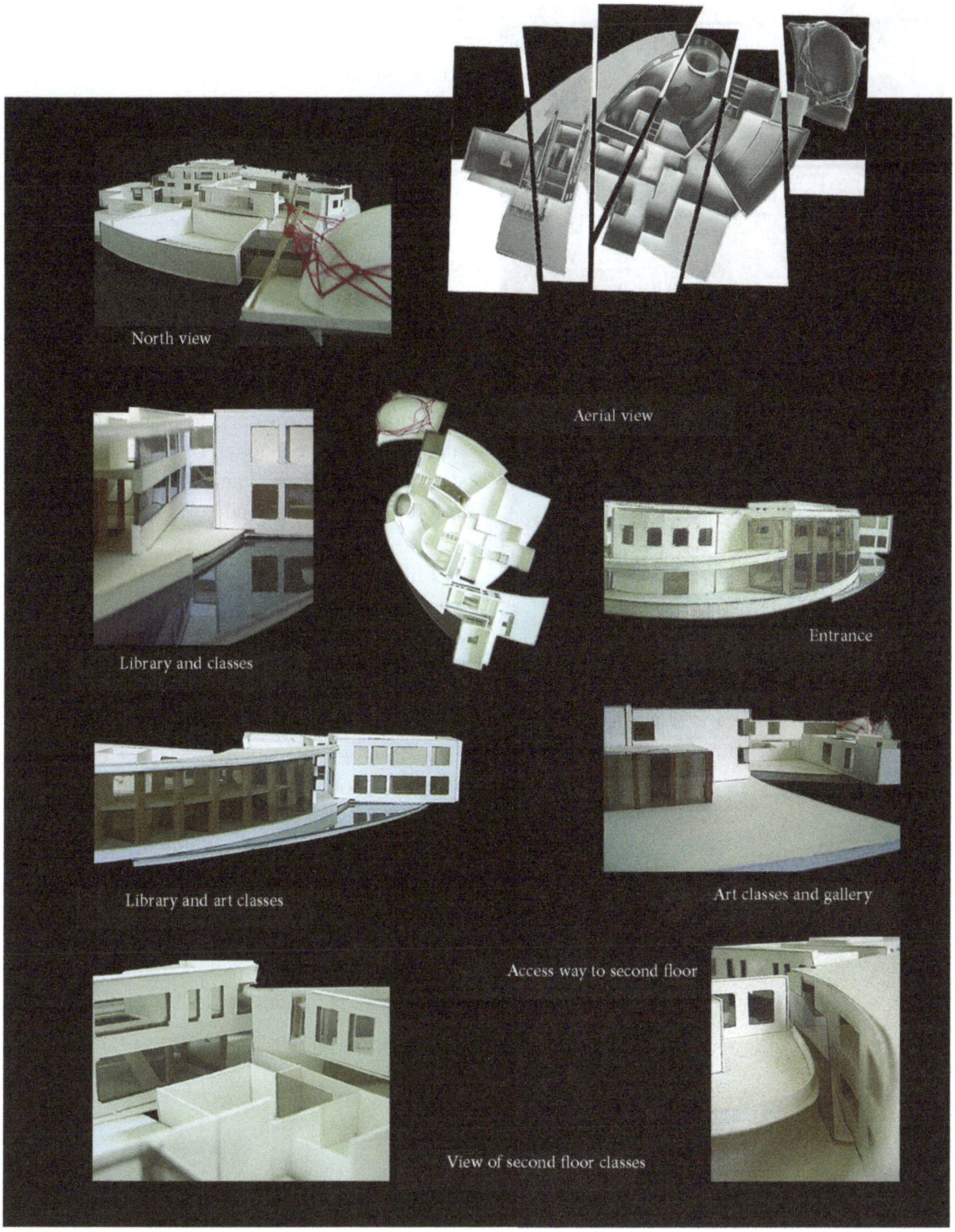
North view
Aerial view
Library and classes
Entrance
Library and art classes
Art classes and gallery
Access way to second floor
View of second floor classes

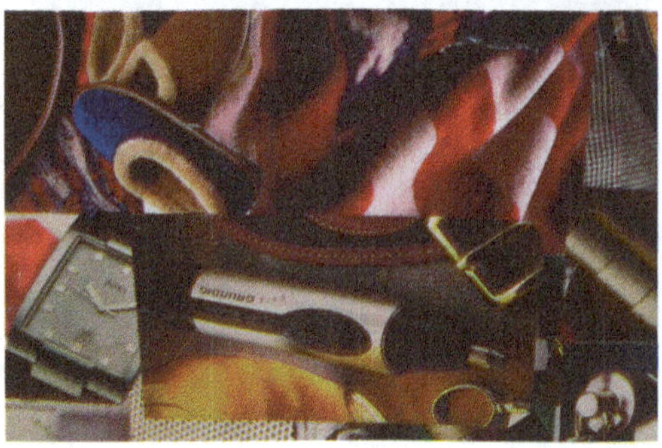

3.30 A Trip to Inside

Chizar cultural center

Student	Shila Ahsaie
Masters	Dr Ali Asghar Adibi
	Dr Ahmad Ali Farzin

Neglected creativity

Starting the design was easier due to previous experiences. I personally accepted and did not stand against it. I thought that understanding of images of collage needs remarkable feeling, but it is not. It is necessary to think ideally and submerge into collages. Design started with studying magazines, finding and cutting images according to patterns and colors.

I like warm colors and ones, with better light, was chosen to show materials obviously. Then, I attached images beside each other, looked them from distance to arrange more. This method is particular and it is impossible to have same collages or spaces. I could see space in collage this time. Bulges and depressions were completely obvious. Colors, beside each other, seemed beautiful and attractive. Unity among spaces could be felt and they represented themselves. I saw an enormous shiny space, planes came out of it and moved around and finally, they diverged. There were spaces among them which differed according their colors, lights and height.

Images seemed explicit; A long plane with light point at the end, A short dead end plane, a string of vertical plane which seemed hanging between ground and sky and go up and down till find an end point. All elements speak and tell their reasons of existence.

There were lots of narrow line lights which showed that open end corridors are too many. A rigid volume could be seen in the corner and seemed to sink into ground. There were dots around volume which should be correlated to it. Gaps were into walls that may wanted to absorb us toward middle of walls, place which ceiling heightened and light penetrated inside suddenly. A narrow line of light, penetrated into one of volumes, was connected and disconnected continuously, it seemed walls disconnected it, but the light hoped to continue its way. All features were mixed and unity could be felt between them. Several rectangles were combined explicitly which their movements provided different spaces and stairs, corridors and main space inside it. Squares formed a repetitive and rhythmic line and its interior was light which none could enter it. It was not appropriate for human or movement but a motionless or static elements such as tree.

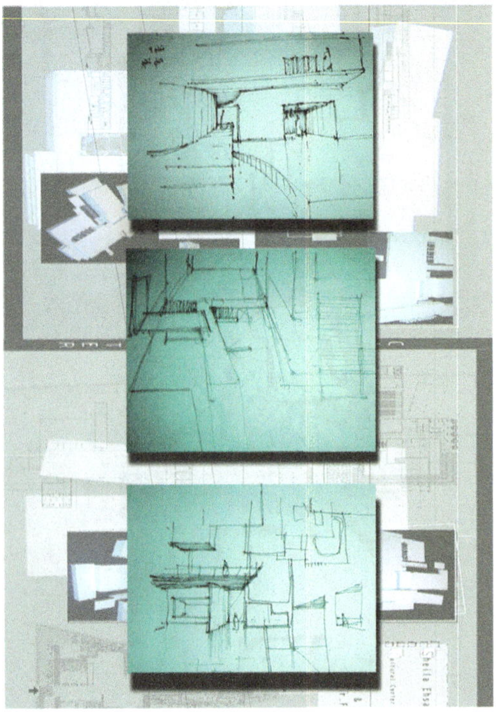

The rhythms separated two vast spaces suddenly. Summarily, these line are bold which represented its dignity and we could stare, move on it and create new spaces. It was time to sketch. I drew line that ascended and descended. Situation of planes, ups and downs are obvious in sketches, but planes with level difference should be connected. Grooves of wall were drown but I should think about it to know its end. It might end to sunken building into ground; or to the combined rectangles; or It might ascend to a bright surface; At first, I drew several columns to create mentioned rhythms but changed them to closed space after a while and drew some trees inside it. The extended line which cut by wall, might be creek passing walls and penetrated into spaces. Some friends sketched my collage, their thought seemed same as me and it showed my spaces were based on reality. All obtained their identities. I started to make a model. At first entrance should be made. I was continuously looking at collages which were like compact disks included all matters. Designing went on progressively. Cutting cardboards and pasting them, I felt to define behavior of people in spaces. Since user came from all masses with different interests, the designing was getting harder gradually. Spaces were being arranged one by one and my desired spaces were formed. I gradually improved my ideal spaces which finally whole the designing was over and different functions were defined in that. "All parts of design belong me", this is advantage point of this method.

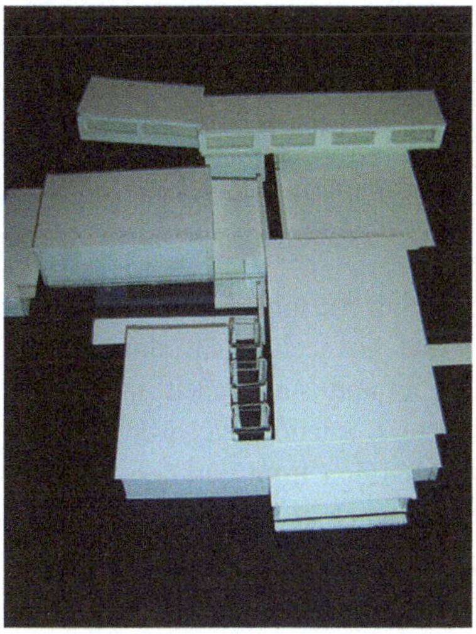

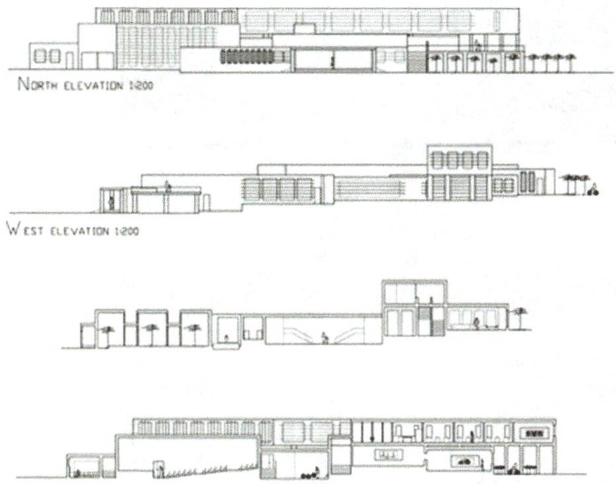
NORTH ELEVATION 1:200
WEST ELEVATION 1:200

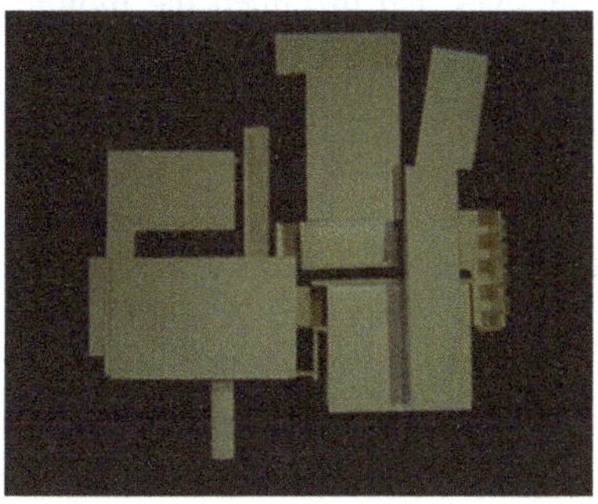

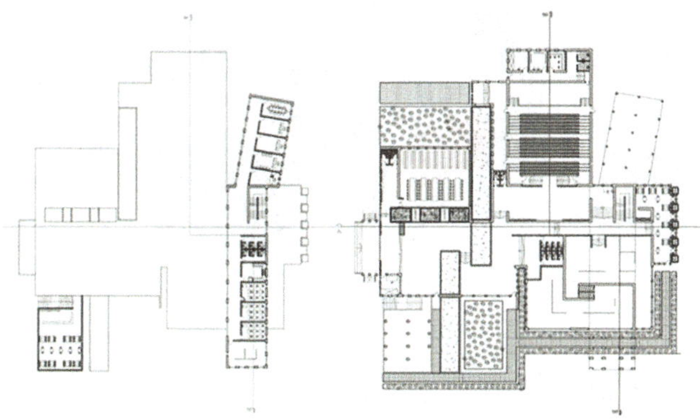

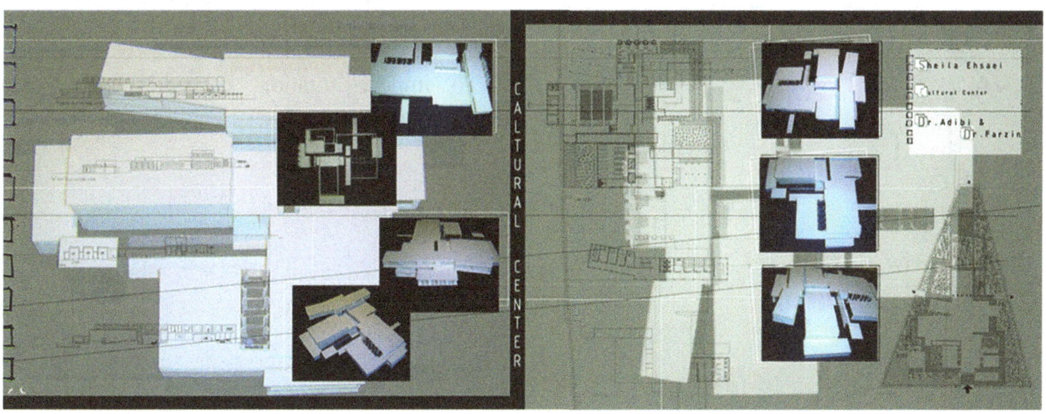
CALTURAL CENTER
Sheila Ehsaei
Dr.Adibi &
Dr.Farzin

3.31 Me and My Feeling, All Alone

Top: Final model
Middle: First collage
Below: Sketch of first collage

Chizar cultural center	
Student	Fatemeh Hashemi
Masters	Dr Ali Asghar Adibi
	Dr Ahmad Ali Farzin

I closed my eyes to sense loneliness before image selection. Me and my feeling, all alone. Peace and quiet. All around surrounded in silence. Feeling of lightness, rotation, etc. were moving down slowly. A motion from darkness to lightness. I sensed presence of brightness gradually. I was moving to static point in wind flowing and light weight, quiet and scented space and sat as soft as feather.

I opened my eyes and imagined the moving direction of feather in space. I looked for images among planes which were responsive to my necessity to lightness, softness, rotation, motion, silence, brightness and shadow. I put all parts together. The whole senses were active. A complicated ribbon was turning point of collage which ascended rotating from ground to a bright point. It had to perform independent role, the first role in space. I chose a image that imposed sense of light reflection, like a fist spread yellowish beam of light and the mentioned ribbon is rotating around it and ascending.

To impose rotation, curved sections and images were put together.

Dark and bright spaces, spaces with different tones of lightness were combined dancingly. It seemed eyes were on lines and planes and sliding on hanging space and correlating spaces one by one and finally, this motion would be ended to silence.

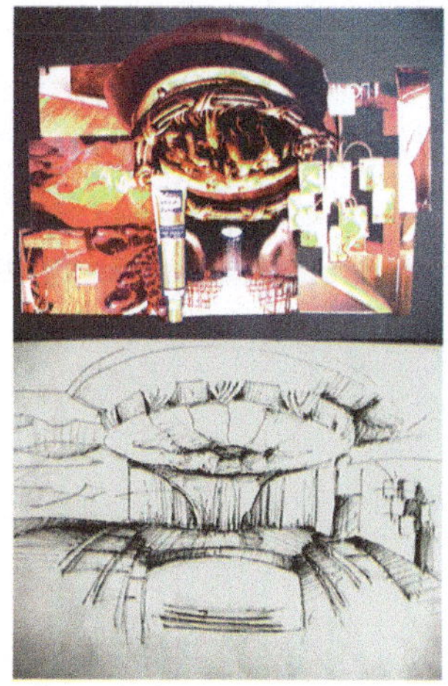

Top: Second collage
Below: Sketch of second collage

When I analyzed spaces of first collage, I found out they were enough to all cultural center functions except amphitheater. Needed lights and shadows for amphitheater could not be seen in collage. Therefore, I decided to make a fresh one or it.

I was interested in light playing in amphitheater. Dark and semi-dark spaces and light strips cutting the volumes. Therefore, I found warm color images and semi-dark ones, based on needed lightness. Parts were put together, absence of something could be felt. I wanted to design special ceiling of amphitheater. An attractive one. I found a element with particular lightening and suitable for other ones. Images was completed in manner that if I changed my view point, I would see entrance and exit doors and all secondary needed spaces. I even could tell where the connection between this collage and first one happened.

Top: Last collage
Below: Sketch of last collage

I tried to create open and semi-open space in final collage. Personages play important role in animating these spaces. The social space were created. All were moving. A space for interaction. The exterior space of first collage was obvious in this collage. A ramp was rotating around curved volume of first collage and passing through ground slope and finally ended to one of cultural center entrance.

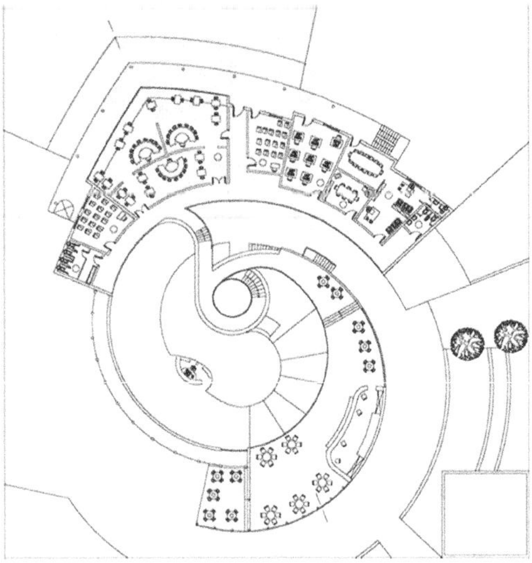

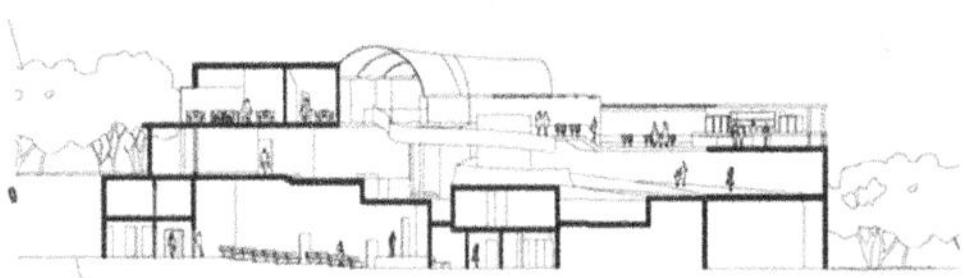

Top: Plan
Below: Section

Prepared the collages, I dealt with sketches. Then plans were provided based on perspective sketches, initial models were made so imagination come true. I used the first collage to design main spaces. At first, I tried to make obvious the vague images of collage. A stone made ramp which sand covered its floor as representative of earthy world rotated around an aquarium, as representative of life in sea. It seemed water donate it life and give to soil so as to bestow it to sky. Ceiling of complex was the blue in this side and seemed that naked water narrate the story of sun, moon and stars on the presence of sky.

I thought the elements should be defined in space at which people spread their wings and travel to imagination world; A space to talk in emotion language. Then, I designed the galleries around imagination core.

It was time to step from emotion to logic world. I dealt with result plan and models and tried to define access ways. It seemed one continued the line on plane using invisible ink and I just colored it. I simultaneously analyzed the needed functions and tried to place them based on spaces of collage. In doing so, library was defined in first floor and official and educational rooms in second floor.

Designing of ceiling and floor were done as best as possible to provide absorbing image of space from everywhere. Whatever one could see in gallery, other could see it from different point as beauty as the one in gallery.

I defined cafeteria in semi-open space up the gallery to provide eye-catching landscape for who is resting. In doing so, I designed other spaces.

Finally, I had to design suitable landscape. Its design had been formed in my mind when I was designing spaces. Site was also based on hierarchy and nature. I provided feeling of rotation and moving toward upside using present motions in site. Trees, fountains and flowers were included in site and at last designed was done.

Top- Middle- Below: Images of model

Right column (top to down):
First collage
Second collage
Third collage
Sketch of dark opening
Left column:
Left corner of first collage, dark opening

3.32 Scattered Ideas

Chizar cultural center
Student Ali Mansouri
Masters Dr Ali Asghar Adibi
Dr Ahmad Ali Farzin

Having problem through timing and collage usage in previous experience, it might cause me to choose other designing method. Instead, pleasure of understanding of architectural space using collage directed me toward it once more. However, there happened problem such as timing again, but result and understanding of project, its architectural space, which achieved at the end, would provide feeling of enjoyment but not regret.

First collaged was created and it approximately resembled to my ideal color and light. Sections and lines of images caused a non-perspective formation of images and no more was suitable to represent a space. The collage possessed the whole main characteristics of spaces. I perceived it as a section of spaces after a while.

Darkness, narrow and low impact lights, at the left side of collage, formed a space in my mind and I named it "Dark hole", which was a dark and heightened space at. There were some access ways passing it in the middle and water flowing, which water motion and its elongation from north to south were considered in site detail. Access way which was in the middle, based on Persian garden, directed toward water source (against direction of water flowing).

Light penetrated through narrow and long groove of south wall and shone in water and its reflection occurred in north wall and ceiling. This space later was defined as intermediate between amphitheater and exterior due to its lightening, silence and calmness.

Dark space, groove of green space and central space, lobby, with its warm colors and natural and superficial light were followed one by one. No more information related to mentioned space was in this collage. Actually, designing of lobby had been started in second collage before it resulted from first one. Teachers considered this space in front of main one, which important matters happened in rear space. There were stairs which supported by a pair of vertical walls and a gallery that light penetrated inside it through groove at the end and its height was adjusted by an upper space. Wall at right side of gallery extended in exterior. There was rotation angle between directions of stair and main space. However, it was omitted due to some reason. Therefore, the whole space directed north-south or east-west. Upper part of gallery which heightened 3 m above ground level, was modified to access way gradually and correlated all spaces. It later appeared to be the most obvious element. The correlation between this path and stair was the prime reason of omitting the rotation angle between direction of stair and other walls of lobby; In doing so, the mentioned path, in its way, had rotation angle related to walls which it was absent due to its extension toward direction of spaces such as dark hole and light hole. Designing the circulation paths of spaces, requirement of gallery to separator in right side wall caused me to define a thick curtain. Entrance was designed based on geometry of this space and its planned facade.

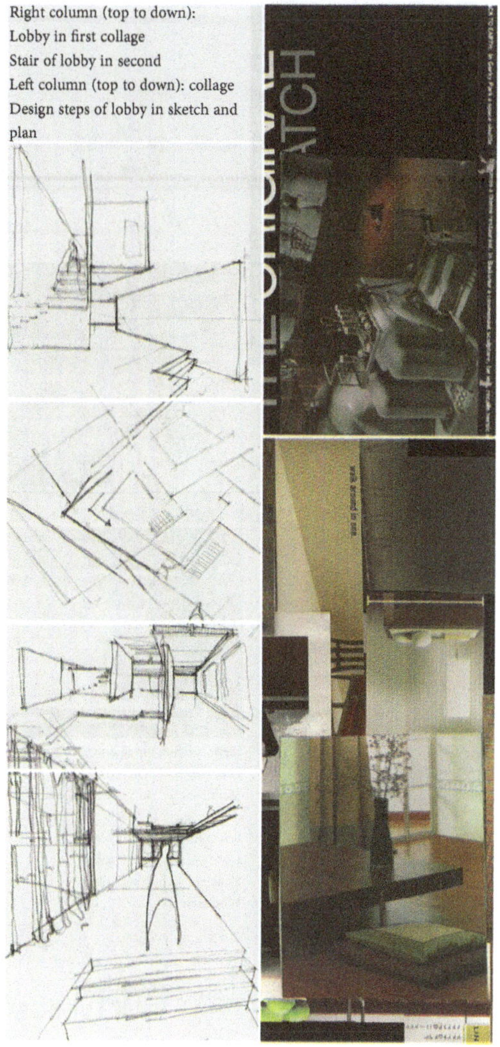

Right column (top to down):
Lobby in first collage
Stair of lobby in second
Left column (top to down): collage
Design steps of lobby in sketch and plan

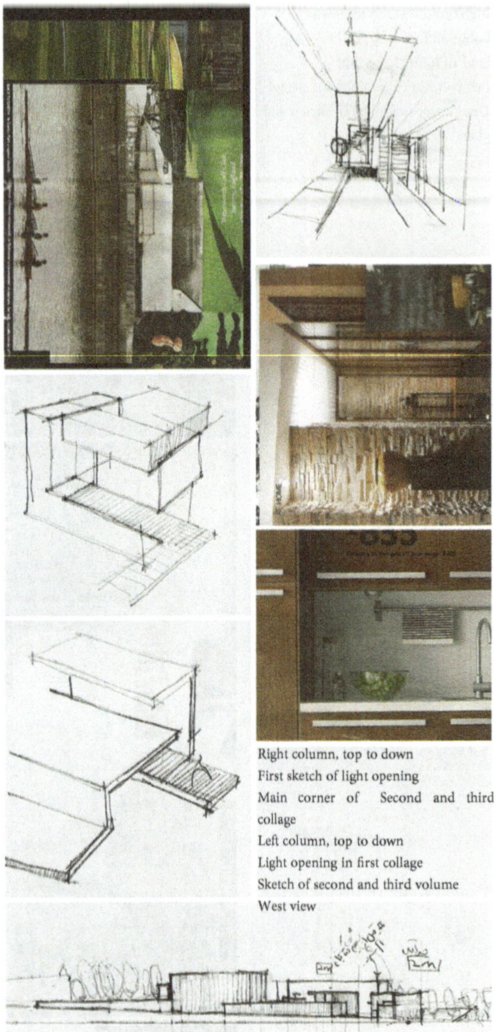

Right column, top to down
First sketch of light opening
Main corner of Second and third collage
Left column, top to down
Light opening in first collage
Sketch of second and third volume
West view

There was a long light place other side of lobby in first collage which named it "light hole". It was semi-open space, at which present Populous trees located, half of it was covered by lawn and it connected office rooms, classes and library with lobby. Green color and light penetrated through transparent wall between lobby and the mentioned space. Passing a space full of light and nature to enter calm spaces including class or library, make people feel peaceful and out of busy world which is necessary to be present in mentioned spaces. Access way to library was separated from one to classes about 3 m above ground level.

Beside third collage was based on previous collage, it included ideas of volume related to combination of exterior and interior spaces, applying landscape and some spaces with warm colors which formed outer shell of open space. Left side of second collage included ideas related to volumes which volume of amphitheater was designed based upon it.

Arranging the scattered ideas and change them into one was the main problem. I tried to provide unity among ideas for a while but I could not. Finally, I decided to represent them separately, instead of combining them. Therefore, access ways which located at 3 m above ground level, passed through spaces. Even this path differ step by step from beginning, in exhibition, to end, in gallery.

Locating spaces in site, connection among spaces and also between them and open space were remained till this step of designing. Group of membranes were defined to make connection between volume and shell of general structure. They ascended from ground and it was possible to donate some information to users.

Generally, design is based on prime characteristic of spaces of first collage including long and separated shells. For color and light of spaces, I tried to apply heat of central space and colors which were based on nature.

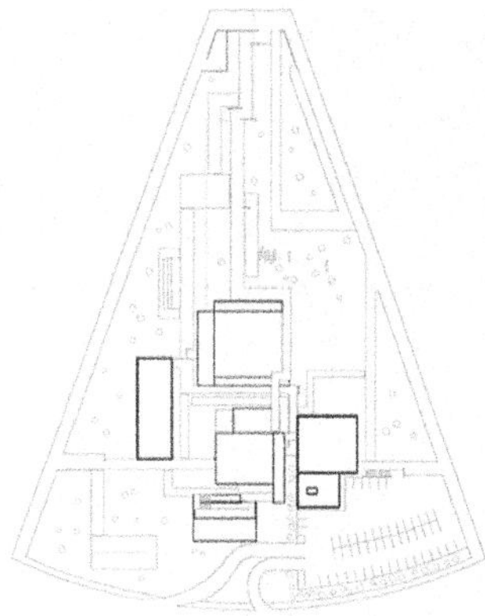

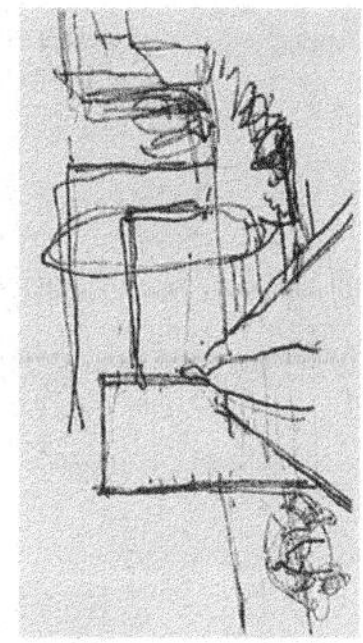

Sketch of east entrance
< Site plan

Sketch of west entrnce
Ground floor >

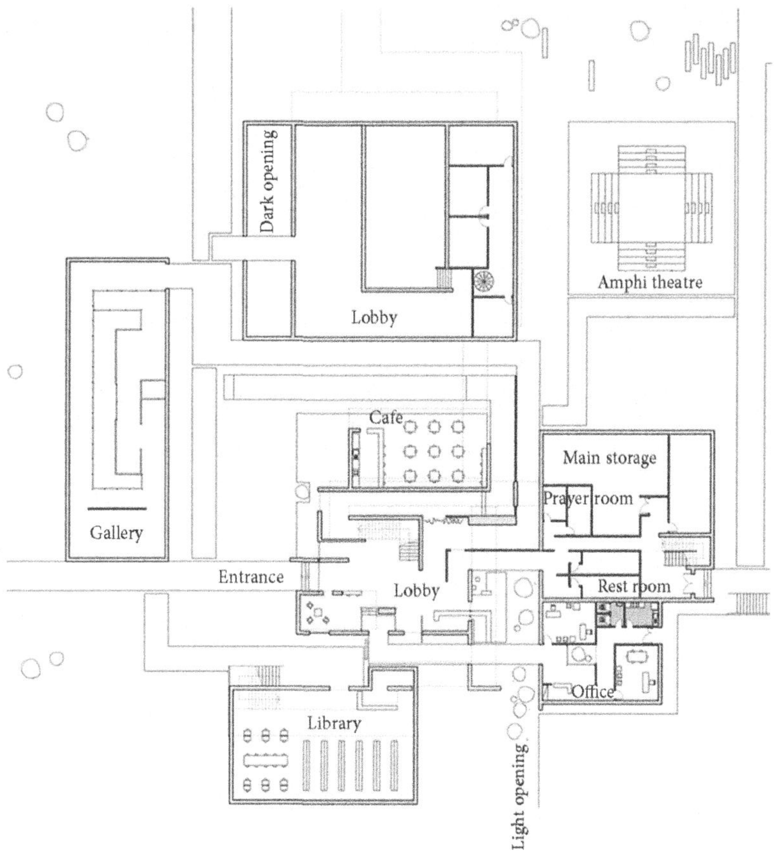

3.33 Creation of Space

Chizar cultural center

Student	Majid Gharibian Lavasani
Masters	Dr Ali Asghar Adibi
	Dr Ahmad Ali Farzin

It is hard to explain where creation will be occurred or it is hidden.

Nature of our witness is particular when we are observing creation of something, therefore explanation of that time is impossible. Prime principle of this method is perhaps process of form and artistic combination of elements, materials and structure. Understanding of this subject needs long time especially one who, same as me, is not familiar with collage.

Collage opens new field of understanding, fresh designing and talking. It is pure of the whole previous guaranteed principles which cause none charging. Creation of new spaces, based on its functions, inform us a fresh life which is changeable and possess different meaning and talk about new matters.

I started designing. At the beginning, I represented my spaces. Walls, light, structure and elements related to technology or full and empty planes were perfect and flawless.

This space was what it should to be, following steps was considered as effort to design it better.

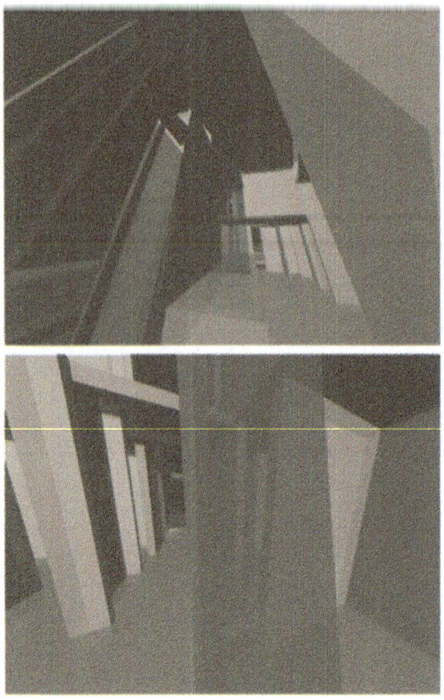

Up: Interior space
Low: Ground floor

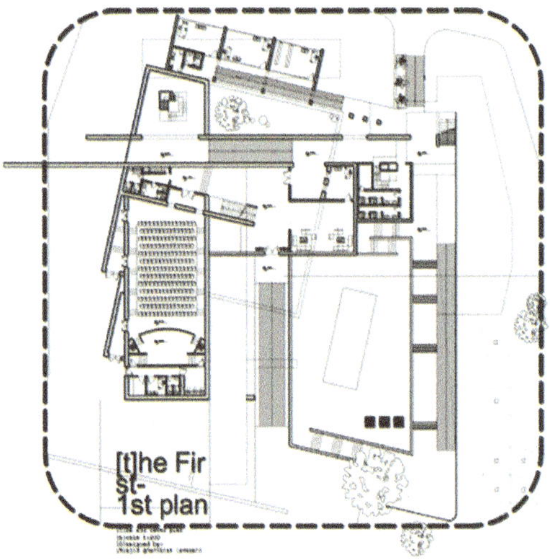

First collage was not suitable, in the other word, none space defined cultural center, It rather than resembled to a barber shop. First collage was not suitable, in the other word, none space defined cultural center, It rather than resembled to a barber shop.

Therefore, I immediately started to make another one. Images and parts were chosen based on form and its material. I used big images of architectural spaces less and was interested in rhythmic lines and colors of them. For instance, when I applied image of airplane, I found lines and bulges of its propeller like ceiling of corridor which its rhythm directed us forward.

Arrangement of images created a composition which was strange, even for me. Two reason caused special lines in abstract of collage: rare, trapezoidal and pyramidal forms of pieces and their strange arrangement.

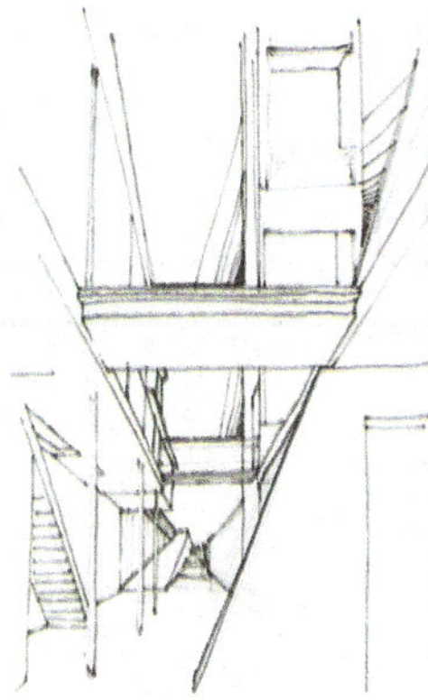

Top, Right: First collage
Top, Left: Sketch of first collage
Down, Right: Sketch of second collage
Down, Left: Second collage

This created duality. Should I design based on sketch or my feeling?

Initial sketches were quick or even incorrect which caused formation of imperfect space but sensible instead of abstract and flawless space. I could get closer to design after some quick sketches. I sketched once more but more detailed and accurate to make a model.

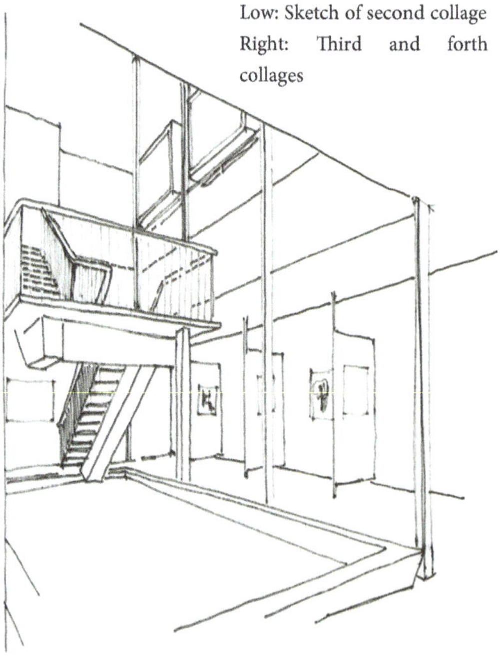

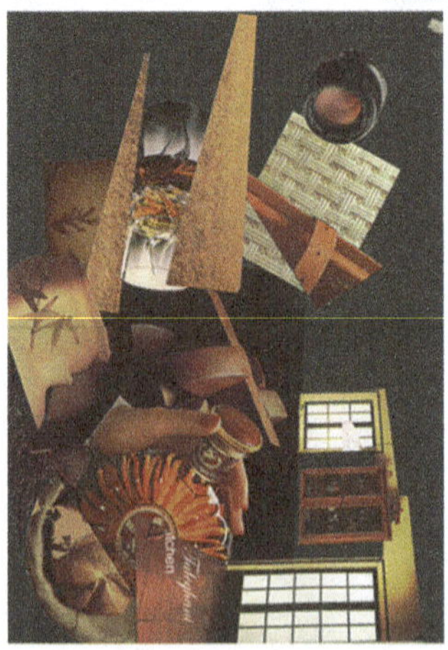

Up: Second Collage
Low: Sketch of second collage
Right: Third and forth collages

Making model was simple or even attractive because I knew model so well and freedom of action by collage. Problems of sketch solved and designing got closer to collage. In this step, using real type and color of materials formed interior spaces. Moreover, facade of cultural center was designed in my mind. It caused to improve rapidly.

Models of lobby, amphitheater and gallery were done and main entrance was defined. I completed geometry of design including in plan and section, So that I could continue the design without collage. However I planned access way among classes, library and computer center. It was no more time to apply this method further.

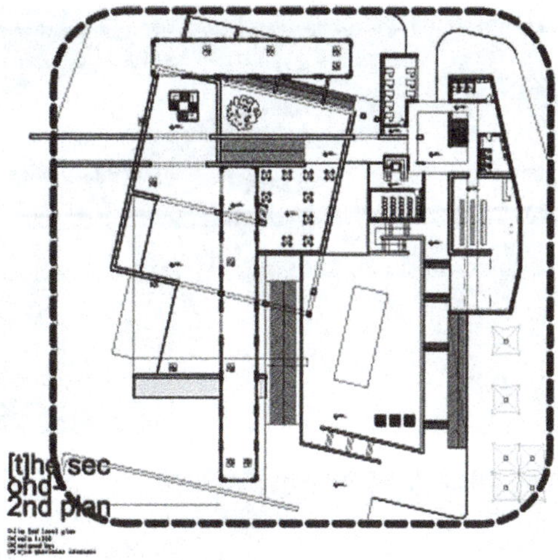

Up: Second floor plan
Low: First floor plan
Left: Sections

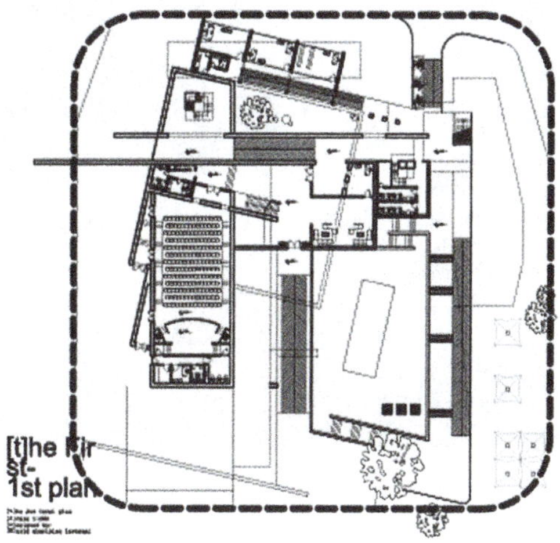

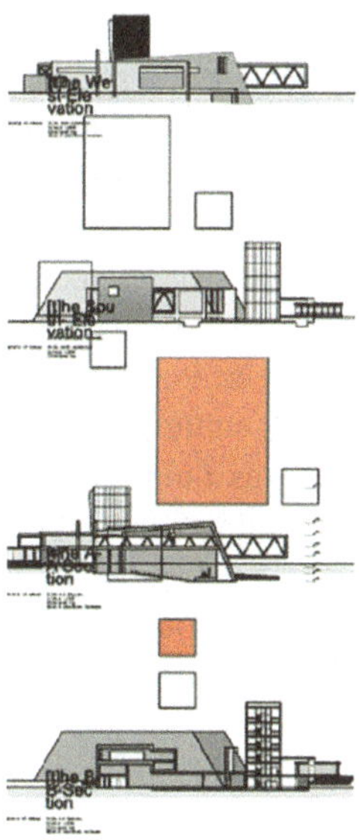

Up: Computer base modelling
Left: First collage

3.34 My Neglected Pictures

Chizar cultural center	
Student	Fatemeh Vahid
Masters	Dr Ali Asghar Adibi
	Dr Ahmad Ali Farzin

I attached the collage on windows of design studio, came back a little to look at it. I realized interesting matter, that bright golden sunshine all around world penetrated into my project (collage). It was totally by accident but showed the space unitedly. I felt amazing because it learnt me how to perceive separated spaces in one through single light and space.

It finally happened!

It was first time which I understood space after efforts! It was wonderful experience. I tried to observed with eyes which help me to see willingly or unwillingly, intentional or unintentional, at last happened. I could move into different levels and arranged spaces continuously and fluidly. Blue-golden, light and transparent spaces. I assume selection of these images was based on effect of culture on my unconscious mind. The space was fluid and rotating. Beside its present internal attraction, it directed your eye outside.

This collage which was close to my ideal mind, I looked for a light behind different barriers and materials. I felt it at the center of collage (interior yard was defined at the center after a while). Created space and access ways directed eyes inside of plan. This kind of flexible and interactive space was certainly effected by culture. I mean that if I designed a educational or art center, selected images would not be like these ones.

Up: Model of collage
Low: Computer base modelling

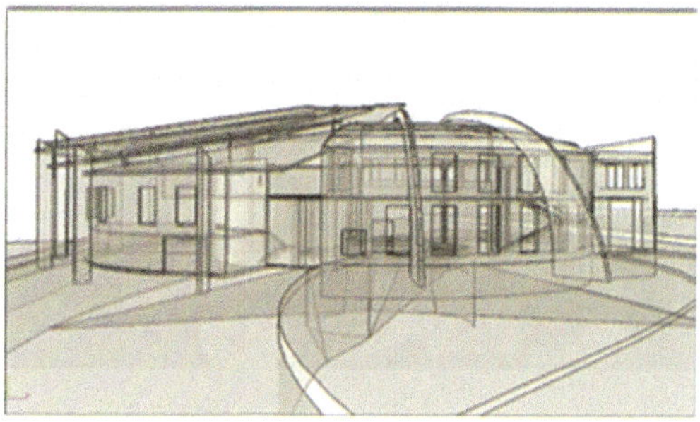

What is position of innovation in this method? Does selected images bind the creative mind?

Is mind achieve of images?

Being alive (notion of space designers (Spatiography) who tell starting point of our method is realistic and alive spaces) or make alive? (It is the matter happened in classic system). Both have common goal and that is "thinking and presenting a valuable design based on architecture principle". The single difference is about starting point. In spatiography, we work according to objective images (ones in front of you) but in classical method based on mental images. Are mental images the ones we are observing in real world? None can claim that there is nothing in real world for each mental image, generally, It means basic elements are same in both (mental image in classical method and objective one in collage). Their difference is obvious, when creative mind start to choose and combine images. Thus, this process is exactly in common (classical) method but its image process is invisible and we just observe final result and sketches effected by it.

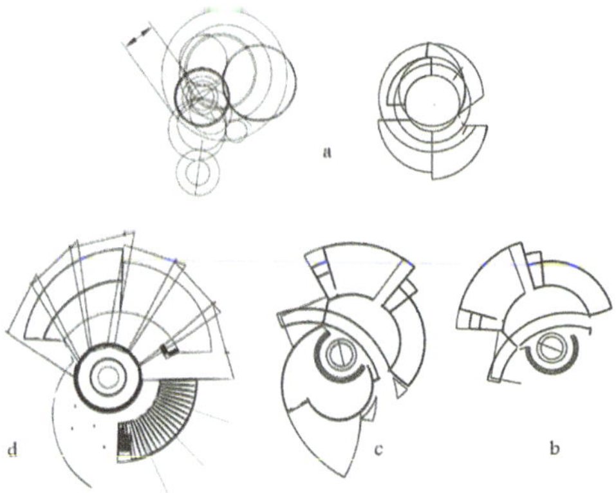

Up: Initial sketch of collage
Left: Model of collage
Low: second collage

One of advantages of collage is that we never forget how we get from a to b, c, d, …. This management make the process easier or more active, because design sometimes gets changed more than usual which even its creator mind gets confused to recognize it. Since first images are considered reference in collage, problems and difficulties in managing the design are less.

Selected the ideal images of magazines and journals, next step is began; arrangement of images (collage elements). I arranged them in manner that space connections are maintained. This connection are caused by preserving continuity in common visual elements such as color, form, etc. Due to this reason, I do not recommend this method to amateur students unlike ideas of opponents of this method (based on their opinion, collage is, because of its suitable for first year of university students for subjects including composition and basic of design) because starting point as reference event, requires personal management of student, also it is individual process which knowledge, inside and conscious mind of students cause it to create and certainly he understands whatever in collage, his teacher understands differently. Therefore, teacher just guide students and imposing personal interest of teacher in student designing is against the principle this process. Therefore, since principle of design is necessary for amateur students, this method is not suitable.

It is assumed that students who choose collage, they know principles of visual arts because these principles is necessary from the beginning (selection and arrangement of images) and managing the project (unconsciously).

As collage encourages creative student but fearful, it also covers inability of amateur in the context of representation and performance.

Last words

Imagine a professional painter draw his works with six colors; this method adds neglected colors to his/her color palette. It means that collage make obvious intentional or unintentional neglected matters to recall his/her forgotten images.

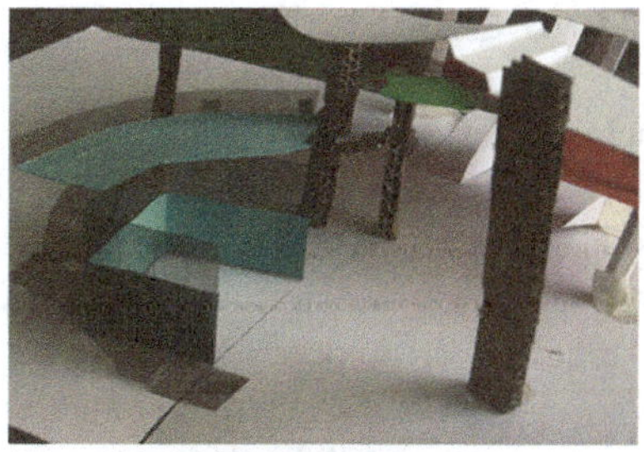

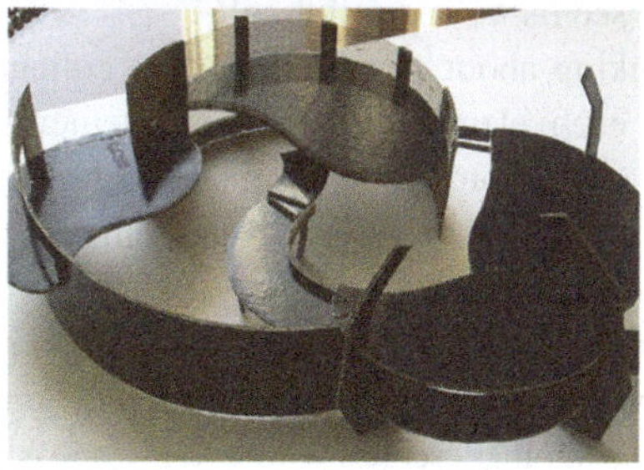

Top to down: Model of collage - second model - Computer base modelling

3.35 Two Case Studies

■ Conclusion of collage process

Actually, changing from collage to three-dimensional space is modification of subjective and abstract phenomenon to objective reality which requires creativity and vast horizon, moreover, since it is considered as natural process, it seems no more difficult.

Thinking about created images in collage, we realize each element includes subelements that appear as a complex and integrated form. In addition, none space is wasted among them, because even distances is part of general.

Moreover, hierarchy of this complex is not completely separated, so that each space through intervene in other create a whole. It is impossible to remove unstable and slippery borders, because these border cause unity of complex. Finally, an integrated complex is provided with no more contradiction; an identified, continual and combined complex as a regular whole which attain particular harmony between regularity and irregularity.

It is a fact this fresh and dynamic identity is formed in a way that allows each subelement of complex correlate with surrounding. Collage which defined as pure of arrogance and responsive to real necessities of images of collage that may solve at the first glimpse, is changed to the most fundamental characteristics of ideal life and these values is result of internal force and hidden memory of designing.

When internal gestures of structure are united regularly and geometrically, its exterior will be arranged. Thereby, each fresh and dynamic structure reach its particular identity which all admire it and removes anxiety of designer, one who was follower or conservationism before.

An explicit and obvious principles make the complex attractive; Principles based on nature, designer interest and law of nature, guides designer to self-knowledge.

Through deeper understanding of images, we realize, if nature of project and necessities of designer is in harmony more, the mentioned identity is more dynamic which its results are formation of reference point and creation of design.

Recognizing his/her nature or generally self-knowledge leads designer to fundamental regularity. In other word, collage never increases knowledge of designer, instead, it recalls neglected knowledge, the one which is hidden behind imposed idea. This fresh and hidden opening of mind provides context to the leakage of designer's internal essence.

It is proved that it is necessary to create images in mind, previously, in order to achieve a dynamic and unity whole which hopeless is its result and it causes deviation and recession in designer's mind. Therefore, it is urged that no images should not form in mind before designing.

Confidence and self-esteem and avoiding anxiety are necessities of this process to forget obvious images of our mind.

At the beginning, ones may refer to that familiar images, because he/she thinks if one forgets them, no more images will maintain. However, based on design subject, site and analyzing, as soon as looking them, they will mark in designer's mind. Moreover, real spaces are created according to these abstract images which increases confidence of designer more.

For one who is not able to avoid these images, because one assumes it as source of information, it is impossible to imagine creativity without that predefined images. But when one believes this and feel comfortable, one realizes that ideal images will be created by looking images well, processing in mind and thinking.

Collage is started from inside and allows images to create interior spaces. In doing so, It direct designers to ideal space.

Naturally designing without precognition is significant matter. Believing the collage, Fear of creation of strange space is neglected and designer feels peace using these method and realizes how his/her mental abilities is limitless and make it possible to combine realities, at the absence of imposing, with mentalities.

Finally, one greatly believes that when he/she understands result of forces, free of any ambition and truth exploring, a dynamic and fresh complex will be created and one can continue the way knowledgeable.

Being loyal to these images, designers gradually feel that they can be free of artificial images imposed by nature and deal with three-dimensional modeling through realizing the direct material of images.

These created spaces and images make person free to recognize him/herself and do and manage anything which is in harmony with his/her essence and nature in order to create a natural thing at the presence of exterior pressure which affect his/her interior feelings.

When designers confront images, actions and reactions will happen. These evolution in thought reminds designer different images of hidden perspective of his/her essence. Later, nature of images are studied and his/her mind forms necessary connection among them through understanding of characteristics and similarities of space and finally, portrays them in objective spaces; Understandable spaces and transferable to other.

References

Betanzos Pint C (2001) Introduction in Bernard Hoesll: collage. University of Tennessee Press, Knoxville, p 1

Bouvier Y, Cousin C (2005) Ronchamp: chapel of light. Besancon, CRDP, Paris, p 12

Cohen. The rational rebel, p 9

Drexler A (1960) Ludwig Mies van der Rohe. Goerge Braziller, New York, p 17

Hejduk J (1971) Education of an architect: the Cooper Union School of Art and Architecture, 1964–1971. Cooper Union, New York, p 280

Holl S (1996) Interwining. Princeton Architectural Press, New York, p 12

Honey S (1986) Mies van der rohe architect and teacher in Germany. In: Achilles R, Harrington K, Myhrum C (eds) Mies van der Rohe: architect as educator. Illinois institute of Technology, Chicago, pp 37–48

Jennifer AE (2014) Shields. Collage and Architecture, Routledge, pp 34, 56

Koolhaas R (1991) The terrifying beauty of the twentieth century (1985). In: Lucan J (ed) OMA-Rem Koolhas: architecture 1970–1990. Princeton Architectural Press, New York, p 155

Koolhas R (2004) McCormick Tribune Campus Center. In: Kiser K (ed) Arcspace.com, Feb 18. https://www.arcspace.com/architects/koolhas/McCormick-Tribune

Le Corbusier (1965) Textes et dessins pour Ronchamp. Association oeuvre de Notre-Dame-du-Haut

Les nouveaux systemes de representation en Architecture 1000 jour pour L'Architecture

Libskind D (1990) Between the lines: extention to the Berlin Museum. Assemblage, pp 18–57

Motahari M (1991) Talim va tarbiate dar Islam, 10th edn. Sadra Publication, Tehran

Rowe C (1987) The provocative facade frontality and contrapposto. In: Le Corbusier architect of the country. Arts Council of Great Britain, London, p 26

The manufacturer's authorised representative in the EU is Springer Nature Customer Service Centre GmbH, Europaplatz 3, 69115 Heidelberg, Germany. If you have any concerns regarding our products, please contact ProductSafety@springernature.com

Printed and bound by CPI Group (UK) Ltd, Croydon, CR0 4YY
30/09/2026
02225251-0020